Bright, Brave, Open Minds

Engaging Young Children
in Math Inquiry

ISBN: 978-0-9776939-8-6 (print)
ISBN: 978-0-9776939-2-4 (ebook)
Library of Congress Control Number: 2015933110

Bright, Brave, Open Minds: Encouraging Young Children
in Math Inquiry, by Julia Brodsky

Illustrations by Ever Salazar
Cover by Sofia Komarova
Edited by Carol Cross, Ray Droujkov
Layout by Petr Savchenko
Copy editing by Karla Lant

Published by Delta Stream Media, an imprint of Natural Math®
309 Silvercliff Trail, Cary, NC 27513

This book came about thanks to parents, teachers, and math circle leaders who tested the activities, gave feedback, and crowdfunded the production of the book. Thank you for your support!

Anonymous
Aimee Levine-Collins
Alan Du
Alina Neuberger
Allison Wilkening
Amy Landau
Amy Sinclair
Andrey Lvovsky
Andrianna Acevedo
Angela Harris
Anna Burago
Anna Ignatov
Arunasri Bhadriraju
Barb Knechtel
Barbara Vasilchenko
Boris Brodsky
Bradley Staeben
Caly Lynberg
Carol Cross
Carrie Schaefer
Cassie McConville
Charles Settles
Chris Swinko
Crystal Johnson
Cynthia Gallup
Cynthia Kahan
Dana Dale
Daniel Barg
Daniela Ganelin
Debra Thorsen
Denise Gaskins
Desiree Foltz
Dmitri Droujkov
Dor Abramson
Drew Fabion
Ed Hsu
Elena Engalychevskaya
Ender Berberian
Eric Zitzewitz

Ever Salazar
Fariha Khan
Giovanna Di Iorio
Herbert Tischler
Hong Liu
Hsiu Chuan Lee
Irene Escobar
Irina Malkin Ondik
Irina McDonald
Iskra Adams
Janet Gerber
Jennifer Mercer
Jennifer Perry
Johanna
Joshua Frye
Julie Wessman
Kamran Melikov
Kathryn Grogg
Keetgi Kogan Steinberg
Kirk Huntsman
Kirstie Jenkins
Kyja Wiburn
Kyle Miller
Lakshmi Maddila
Leanne Paulin
Lindsey Sutton
Lisa Huntoon
Lisa Jobe
Lisa McCarville
Lisfette Nazario
Louise Ryan
Makeda Greene
Maria Droujkova
Mariana Hopkins
Marie Brodsky
Mark Lukin
Mark Saul
Marta Kalinowska

Marylin Tebor Shaw
Megan Kucik
Michael Condor
Michael Faughn
Michal Rotberg
Michele Nielsen
Mick Weiss
Monica Utsey
Myra Haase
Nadezhda Plotnikova
Nichole Middleton
Nina Vaykhanskaya
Olga Radko
Pamela Copple
Paul Gouin
Phonphum Humphrey
Radhika Ramakrishnan
Raj Shah
Ray Droujkov
Remy Poon
Rodi Steinig
Sarah Gray
Sarah Sands
Sasha Chizhik
Shannon Allan
Sheryl Morris
Stephanie Surles
Sue VanHattum
Sylvia Hoch
Tiffany Reevior
Trevor Lim
Victoria Gehring
Ya He
Yelena McManaman
Zipporah Bird

Acknowledgement

I always imagined book authors as hermits, hiding behind their desks and scribbling, humming to themselves. This book, however, turned out as a huge community effort. Literally hundreds of people helped in various ways — bringing their children to our math circles, teaching the pilot lessons, reviewing the text, crowdfunding, and spreading the word — and I am extremely grateful to all of you.

This book would never have happened if not for my husband, Boris, who gave me the idea to start a math circle at the first place, and my dear friend, Maria Droujkova, who offered to publish the book and supported me all the way, from the first draft to the publication. Marilyn Tebor Shaw volunteered her time and expertise to review the legalities with me. And our math circle would not function without our bright and enthusiastic teachers-in-training: Daniela Ganelin, Alan Du, Drew Fabion, Daniel Barg, and Joshua Frye, to whom I wish all the best in their future endeavours.

I had great delight in discussing the book with seasoned educators and math circle leaders, including Anna Burago, Olga Radko, Ender Berberian, Sue VanHattum, and Anna Ignatov, who provided many insightful comments and suggestions. Angela Harris and Arunasri Bhadriraju volunteered to be the first pilot teachers of the course, and provided me with invaluable feedback. Cynthia Kahan let me pilot-teach the course in her all-girls math classroom, so I could see whether the girls would enjoy this kind of math. Ray Droujkov, Alina Neuberger, Karla Lant, and Carol Cross plowed through my English, making the book actually readable. Ever Salazar and Sofia Komarova made crisp, clear illustrations.

And finally, I am very grateful to all the students and parents of our math circle, all the pilot teachers and participants of our "Problem-solving kaleidoscope" event who sent me their questions and comments, and all the people who helped to raise money and spread the word.

Bright, Brave, Open Minds:
Engaging Young Children in Math Inquiry

Foreword

*"Is it possible to educate one differently?
Educate in the real sense of the word;
not to transmit from the teachers to
the students some information about
mathematics or history or geography, but
in the very instruction of these subjects to
bring about a change in your own mind."*
— J. Krishnamurti

I will share with you what I have learned as I tinkered with teaching problem solving to curious young children, ages six to eight. The purpose of this book is to invite you to experiment with your own children or students, without any preconceived notions of how the outcome will look. Instead, allow your personal taste and the children's feedback to guide you.

Leaving well-known ground to explore uncharted territory can be scary. It is encouraging to have at least an idea of what may lie ahead. The eight lessons I share with you will provide some ideas on how to start and what to expect in the beginning. From that point on, you may feel more empowered to chart your own course.

This book aims to introduce beginning problem solving skills of to both children and the adults who teach them. I strive to show you the benefits such explorations can bring both your children and you. I believe you will enjoy being able to face the unknown, whether it takes the form of a challenging math problem or a new way of teaching and learning with your child. I hope that after joining me in this endeavor you will reflect on your own needs as well as those of your children and students, and your future educational journey will gain more meaning and joy.

Young children naturally approach each problem as unique and attempt to play with them. Rather than prescribing specific tools for them to use, aim to preserve their divergent thinking and open-mindedness. "Aha!" problems sound easy, but require surprisingly insightful solutions. Solving such problems enables children to look for elegant, unexpected solutions in other life situations as well.

The solving of a challenging problem naturally takes time and creativity. Our goal is to guide all math circle participants, students and teachers alike, to welcome the time it takes to arrive at a solution.

I highly recommend creating an informal math circle of four to eight participants, optimally; together you can enjoy this book. Social problem solving allows children to present their points of view to a group, defend their ideas, and foster special types of intellectual friendship.

Why problem solving?

Problem solving, the cornerstone of every mathematical discipline, is not easy to master. Problem solving cultivates open-mindedness, the fluency required by divergent thinking, and teaches dealing with fears and obstacles. In order to learn this confidence, children need a safe and supportive environment and sympathetic adults to model how to deal with making mistakes and being stuck. That's what math circles are about.

Unlike the drills children routinely do in school, problem solving requires higher-order cognitive skills. The first is learning to identify when a situation is a problem. The ability to formulate the problem in meaningful terms is another crucial skill. To acquire and hone these higher-order skills, children should tackle ill-defined problems — common in life but rare in textbooks — and learn how to define them constructively.

Problems with easy solutions do not meet our purpose. The "insight" problems collected in this book are all within the capabilities of young children, but most solutions are not obvious at first. My wish is that my children and students recognize and appreciate the feeling of facing the unknown and of being stuck. I hope to give them tools to deal with their frustration, to become aware of their own psychological and physiological responses to the feeling of "not knowing," so that they are better equipped to face life problems later on.

Many problems feel difficult because they force us to discard our unstated assumptions; however, as soon as we do the solution becomes clear. It was very encouraging to me to see that after our sessions, the children were much more willing to consider unusual and elegant approaches to a problem, rather than using brute force. This is the gift of an open mind.

Mistakes have a fearful connotation for children — but turn them into a "sanity check" game, and mistake recognition becomes a welcomed activity. This is a very good skill to introduce at this age, and children burst into laughter when they discover that the solution brings them to a nonsensical result.

Insight problems: starting young

I made an effort to select problems that simultaneously challenge and engage children. I was amazed to observe that young kids tend to favor problems that they don't know how to approach. They readily set out to explore and share their solutions with friends and family. Children especially welcome insight problems. Those problems stimulate creativity and fantasy, teach them to re-evaluate their approach to a problem, and motivate them to search for new ways of solving both counter-intuitive and straightforward problems.

Because insight problems are considered more difficult, they are usually introduced much later down the road, in late middle or high school. I have not encountered much rigorous material of this type targeted at elementary school students. Parents may initially consider these problems too difficult even for themselves, and would not think of presenting them to their children. I argue that six- to ten-year-olds are at exactly the right age to approach these types of problems. More importantly, insight problems boost their motivation to learn.

Summary

My experience convinced me that it is never too early to introduce the joy of problem solving, an enjoyable, educational, and inspiring endeavor for children, parents, and teachers. In a time when our society requires out-of-the-box thinkers more than ever, children need to search for and recognize the feeling of "insight" moments, to distinguish between knowledge and understanding, to allow themselves to make mistakes, to learn how to approach new problems, and to become familiar with their own mental processes.

The most important thing I urge you to do as a parent is to continuously observe your child's development and create a feedback loop to adjust your approach as needed.

> "...For advice is a dangerous gift, even
> from the wise to the wise, and all courses
> may run ill."
> — J. R. R. Tolkien

My math circle values

> "I err, therefore I am"
> — St. Augustin

> "If I had to make a general rule for living
> and working with children, it might be this:
> be wary of saying or doing anything to a
> child that you would not do to another
> adult, whose good opinion and affection
> you valued."
> — John Holt

I envision math circle to be, first and foremost, a hub of free thinking, a place where children can develop without external pressures like fear, coercion, and adult dominance. The purpose of my math circles is not to help students reach certain academic goals, but to let them open their minds to complex problems with deep concepts that they don't know how to approach. I believe that a protective psychological environment that fosters well-being is a necessary requirement for a beginner to open up to such experiences. I would like to emphasize some values below:

Non-coercion

Luckily, the math circle is a leisurely and optional activity. No children should be forced to participate if they do not want to. Coercion of this sort affects motivation and makes the child less able to think openly in the long run. If a child is hesitant, you may invite them to sit and observe quietly for as long as they want. Forcing a child to participate is counterproductive for the child, and threatening for other participants.

Freedom of argument and equal respect toward each participant

A safe psychological environment is extremely important while exploring new things, which are quite intimidating on their own. Fear of authority can completely block the ability to participate in an intellectual activity. Kids should be encouraged to freely debate with their peers as well as teachers, without being reprimanded. Don't establish yourself as an all-knowing authority. This can limit you to teaching your own point of view, rather than developing students' curiosity and critical-thinking skills. The best approach is to patiently guide the kids without drawing attention to yourself, so that they see their own mistakes and learn to draw their own conclusions.

Active listening

While talking to your students, focus on the meaning and significance of what they are saying. Try to see their perspective, to comprehend not only what the person says and does, but what it's really all about. Help them to verbalize their intuitions, and show them the interconnection between their ideas and those of others. Never reject or trivialize what they are saying.

Treasure mistakes

Nothing gives you more information about a child's mental process and growth than their mistakes. Mistakes are gems; lovingly collect and carefully study them, rather than frowning upon them. By learning from the mistakes a child makes, you can modify your teaching to address any shortcomings, strengthen the emergent understanding, and deepen your personal contact with the child.

Often there are valid reasons behind student confusion. The teacher should make an effort to recognize the root of the mistake, but never rush to resolve it for students. Instead, provide gentle hints such as counter examples or taking the idea to extremes.

Making, noticing, and resolving mistakes is a fundamental skill. Use every opportunity to model making mistakes in front of the kids. Hearing you think out loud and discover your logical fallacies will boost their confidence in their own cognitive abilities.

It is extraordinarily fun to collect, curate, and display the treasure trove of the mistakes made by your students. Discuss them in a year or two — children will have a blast marveling at their former understanding. Do it frequently enough and your students will start to see their own progress, as well as what to make of a "feeling of knowing" and "complete understanding" of ideas.

Being stuck

Parents often wonder whether to interfere or not when a child is stuck with a problem, and feel a natural inclination to give a child a hand. However, being stuck is vital to the development of solid problem-solving skills. The feeling of being stuck activates the contemplative and inventive part of our brain. What questions can you ask to help your child? Here are a few examples:

- Can you reduce the problem to a simpler one?

- Is there an alternative way of solving the problem?

- Can you find an analogy in another field?

- Have you checked for hidden assumptions?

- Can you generalize, estimate, build a model, or draw a diagram?

- How would you explain this problem to someone less experienced?

It is important to teach a child how to relax while being stuck, and how to develop an alternative attacks for the problem. Modeling being stuck is also a great way to help a child to relax and see it as a natural process. For more ideas, see "Being Stuck" in Appendix B.

Keeping the mystery

When you see a student struggling with a problem, you may want to give a hint or to ask a guiding question. Curiously, the traditional method of starting simple and gradually increasing complexity may kill interest in problem solving, because it takes all the mystery out of the solution process. Mystery seems to be an integral part of student motivation. Refrain from giving the answer. The child will continue thinking on the problem, maybe for years to come, deepening their understanding every time they come back to it. It is not necessary for you to know the solution to the problem, either. Sometimes, it is more enticing for the student to know that the teacher is struggling with the problem as well.

Secrets of the trade

Thinking time

Introduce students to the idea of thinking time (also called *wait time* in our lessons). Taking one's time to ponder a problem is essential for the development of deep thinking. In math, "quality time" also means "a long time."

Students can only give answers when the thinking time is over. Students may enjoy using a big sandglass or a timer. Remind those who rush to answer before the time is over to capture their thoughts (write, doodle, build). This way, you teach the students to focus on the problem, and to take time to consider alternative solutions. They also learn how to deal with impatience. Teachers also get waiting time just like their students. During waiting time teachers observe their own reactions, take mental notes on student behavior, and decide how and in what order to discuss the answers.

Parent frustrations

It is not unusual for a parent to be frustrated by a child's overall progress or a response to a particular problem. First, acknowledge that you are the one who is frustrated. Own the problem; try to recognize the root cause of your frustrations. They usually stem from your expectations and assumptions about how your child should respond and behave. Next you can analyze your assumptions, question their validity, change them as needed, and think of constructive steps to better support your child.

Making predictions

Make sure that you write down your hypotheses about how the students will react to a given problem, and then check whether you were right or wrong. This way, you will start noticing patterns in your students' thinking and your own expectations.

Not providing an answer

Refrain from providing an answer. It is acceptable for the child to leave the class with an unresolved problem. Answers may be helpful when the child learns how to solve standard exercises, but this book focuses on very different skills.

Multiple paths

Start teaching multiple approaches to solving problems at the earliest age. Even a very young child can compare and contrast different ways of solving a problem. One of the many possible ways to approach this is to ask a child to solve a similar problem and point out the differences between the two. Children develop the ability to solve a problem before they develop the ability to verbalize the solution. There are multiple ways a child can show and model their way of thinking beyond words.

Multiple solutions

Open-ended problems have a variety of answers. Recognizing that there is no single right answer is a powerful antidote to seeing the world in black and white. Do not interrupt arguments about solutions. Write down all sides of the argument where everyone can see clearly. Group discussions of open-ended problems cultivate tolerance of other people's views, and reveal new facets of a problem.

Incomplete solutions

Expect students to forget parts of their solutions as they try to formulate their answers. This is normal, and does not mean that the student has not solved the problem. Listen carefully to students as they explain the solution. Ask for diagrams or pictures that illustrate the solution — this may help the student to remember the missing parts.

The ill-defined problem

The ill-defined problem addresses complex issues and is described in a vague or incomplete manner. While most textbook problems are well defined, real-life problems are not. How do we teach our children to deal with them? Ill-defined problems in this book prompt students to recognize the vagueness, to analyze the incomplete or contradictory data, and to reformulate the problem.

No answer?

Most students do not believe that there are problems without answers. Give them some. Young kids can be amazingly persistent at fiddling with an unsolvable problem. Let them play, even if you know that there is no solution.

The game of questions

New students may be shy to ask questions about a problem. One possible approach is to challenge the students to ask as many questions as they can about the problem in a given amount of time. Afterwards, the group may review the questions and select the most helpful ones.

New notations

Sometimes students invent their own notation. You want to notice, capture, and collect notations, as well as discuss them with students. New notations may open a door to a completely new area of math or science. Picture how Arabic numbers changed the way we deal with arithmetic. Just imagine multiplying Roman numerals!

Introduction of manipulatives

Don't ever start your activity before children have had a chance to play with any new object. This helps children explore the properties of the object, and get comfortable with it.

Definitions

Young kids are unlikely to give formal definitions. If they struggle coming up with a definition, invite them to think of examples first.

Taking turns

Taking turns addresses several dangers. First, it reduces cacophony and chaos. More subtly, even in an orderly group there is a tendency for a few people to dominate the discussion. Use any toy (e.g., toy microphone, teddy bear, etc), and explain that only the person who holds the toy can talk; the rest of the group listens. When the person is done talking, the toy goes to the next person.

Topics

I present here a few topics that I taught at our mathematical circle for six-to-eight-year-old children. The problems selected reflect my personal interests and tastes, and do not follow any specific plan or methodology. They represent the problems in an order that worked well in my classes. I encourage the reader to modify or reorder the problems any way that suits your own needs. As you work with the book, ask yourself: "What makes this problem interesting? How could I modify it to make it more interesting? Less interesting?" In case you're wondering, making a problem less interesting is a way to sort aspects of problems and classify them as exciting or boring.

You may want to supplement the topics with your own resources, including YouTube videos, drawings, puzzles, etc., and share them with other math circle leaders.

Topic name	Math and problem-solving objectives
Introduction to thinking	Diversity in problems; multiple solutions; hidden assumptions
Strange Statements	True/false statements; syllogisms; paradoxes; divergent thinking and analysis of problems and solutions
Tiling Puzzles	Symmetry; tiling; reflections, rotations, translations; verbalizing visual-spatial concepts
Problem Debates	Math communication; elaboration; defending your idea; team work
Infinity	Recursion; cycles; using problem solving to explore a topic in depth
Symmetry	Symmetry in life, science, and the arts; use and misuse of our brains' preference for symmetry
Lateral Solutions	Start at the end and work backwards; use and misuse of common structures and analogies in series of problems
Game Festival	Math magic and party tricks; problem solving in board games; community building

Introduction to thinking

Introduction

Goals Problem diversity; multiple solutions; hidden assumptions; diagrams

Materials

Toy for "microphone"; whiteboard with markers or a projector for recording group ideas; poster, graph, and regular paper; colored pencils or markers; scissors; matchsticks or craft sticks; puzzle print-outs; poker chips or counters; supply of foam animal stickers

Set – Up

Spend some quiet time preparing the room, arranging supplies and otherwise establishing the space. Greet parents and students and help them make name tags and decorate them with math pictures. This invites creativity from the beginning.

> *"The defining trait of our culture — we are talking about kids, but we are not talking with them"*
> — Peter Senge

> *"There's no sense in being precise when you don't even know what you're talking about."* — John von Neumann

The Name Game

Warm - Up

Materials Toy microphone; whiteboard, markers

Set - Up Classroom rug

Introduce yourself. Say something funny, silly, or interesting about yourself, something kids can relate to.

Example

"Hello, I am Maria. I really love space travel, and I just finished making up a game about it." Or, "Hi, I am Joseph. I am a human being. I love engineering and cooking. My favorite game is chess, and my favorite food is ice cream."

Invite the kids to the rug or other "circle time" space in your room. Take a toy microphone or any object you will use as a pretend "microphone." Ask students to take turns saying their names into the microphone, and naming something that they love to think about.

Teacher's Notes

This activity helps students closely relate to one another. It also establishes their central role in the club, as people who bring in wonderful ideas.

Capture answers on the board or a file you project on the wall, or ask the parent volunteer to do that. Ask others to wave instead of saying, "Me too!" to prevent extra noise.

Student reactions

Students are likely to say they like to think about favorite foods, pets, friends and family, or hobbies such as playing soccer. They also frequently name books, movies, or computer games that they love.

Sometimes students wonder if they are only allowed to talk about mathematics: "Are you asking if we like addition or subtraction better?" This is a good opportunity to mention that mathematics or problem solving is everywhere, so they can talk about anything at all that they love.

Discussion # Problem solving

Materials Whiteboard, markers

Set - Up Classroom rug; group work on whiteboard

Wait time Two to three minutes

Tell the students: "Several million years ago, there lived ferocious saber-toothed tigers with sharp teeth, crocodiles with awful jaws, and the first cave people, who had no strong jaws, long teeth, or sharp claws. What helped those early humans to survive in their unfriendly world full of dangers?"

Write down student responses, and focus their attention on the answers that are related to thinking/problem solving.

Say "In this course, we are going to solve strange and unusual problems, to obtain problem-solving superpowers"

Teacher's Notes

The goal of this activity is to focus students' attention, define problem-solving as a basic skill, and connect it to human survival. Do it in a relaxed and light-hearted way. Invite students to suggest examples of situations where they need to solve problems. You may want to remind them that there are no grades or assessments, and that all mistakes are fine. Those kids who may have trouble writing may draw instead.

Hints for the Teacher

After the students are done, read aloud all examples, and confirm that they are relevant.

Students often give specific situational examples. Summarize the higher level aspects of thinking that the examples illustrate, such as looking at connections, or paying attention to details.

Congratulate kids on creating the definition of problem solving together. It may not always be obvious to children what they just did.

Student reactions

Students quickly come up with various examples of thinking and problem-solving situations. They also mention inventions, such as use of sticks and fire.

Discussion **Problem-solving roadblocks**

Materials Poster paper, markers

Set - Up Classroom rug; group work on poster paper

Wait time Two to three minutes

Tell the students: "Sometimes, our difficulty is not caused by the problem itself. We forget to have breakfast, or don't get a good night's sleep. Let's brainstorm together what helps and prevents problem solving."

Draw the "stock and flow" diagram on poster paper. Explain to the children that the box for "problem solving" is like a water tank; that takes some time to fill. Ask the students to contribute to the "in" and "out" flows. Then, ask the students to observe their problem-solving abilities as the class progresses.

"Stock and flow" diagram example:

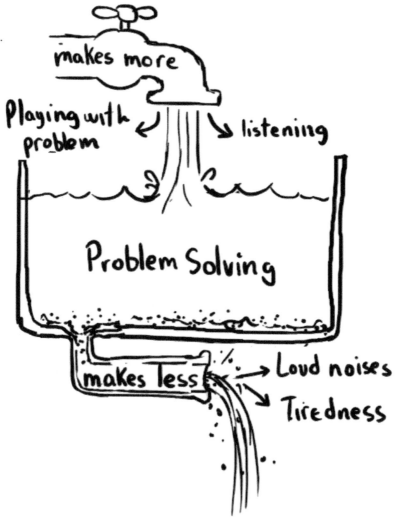

Teacher's Notes

Explain to the students that the goal of our classes is to
make problem solving as deep as possible, minimizing

the out flow, and maximizing the in flow. Ask students to brainstorm various behaviors, figure out where they go, and add them to the diagram.

Let the students know that the diagram will live in class, and they can always add new suggestions to it. Take photos of the poster, and email it to the parents after the class. Parents may want to have a similar poster at home.

This activity invites students to observe themselves as scientists and to look at the big picture behind their actions. On a technical level, this activity introduces students to a stock/flow diagram (one of the basic diagrams in system thinking), establishing the basis for classroom rules.

Student reactions

Students love to contribute to an open-ended problem like this, and see their input on the paper.

Count the Squares

Activity

Materials Whiteboard, markers; graph paper

Set - Up Students at their tables

Waiting time Two to three minutes

Ask students if they know how to count — you will get a resounding "Yes!" Next, ask them to count the squares in this diagram you draw on the board:

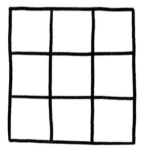

Tell the students that this is a tricky problem, and ask them to think more.

Teacher's Notes

Do not rush the students. Be quiet, relaxed, and attentive as they think for as long as they want. Take notes. Acknowledge and share every student's idea that comes up, right or wrong.

Enthusiastically celebrate students that notice that the problem never mentions the size of the squares. You can pretend-play that you simply can't believe the amazing idea you just heard. Ask students to explain their incredible thinking.

Bring each new notation to everybody's attention. Discuss the pros and cons of each notation.

This activity focuses on problem statement analysis, unstated assumptions, and new kinds of notation.

Student reactions

Expect "Nine" as the first answer from the students. Next, you will probably get a student who thinks of the big square and says there are ten altogether.

Encourage students to copy squares to their papers and to use colored markers to solve the problem. Students may invent new notations: mark centers of squares with different colors, highlight the vertices, or color in their squares.

(Consider the importance of notation — for example, compare doing multiplication in Roman numerals vs. Arabic numerals.)

Solution

9 (1x1 squares) + 1 (3x3 square) + 4 (2x2 squares) = 14

Activity **The Game of Nim**

Materials Poker counters or other small objects, such as grapes, raisins, or Cheerios

Set - Up Students at their tables, working in pairs

Invite students to pair up with a friend. While they separate into pairs, tell a little story:

"An evil wizard has challenged you to a wizard game. He gave you twelve enchanted grapes. You may eat (or remove if you use counters) one to five grapes at each turn. The last grape is poisoned, so don't eat it! Whoever has to eat the last grape loses."

Teacher's Notes

If students are experiencing difficulties, you may want to start with smaller number of grapes, say nine.

For younger or non-competitive children, who may get anxious if they lose, suggest role-playing a more neutral situation (such as two panda bears trying to avoid the last wilted bamboo shoot) so that they do not take it as a personal loss.

This activity promotes seeing patterns, solving from the end, and working with a partner.

Student reactions

Students get into the game enthusiastically. The first few games usually have no strategy, as students play randomly and observe what happens. After a few rounds, they start noticing patterns. Backwards thinking is very counter intuitive for children this age and challenges their thinking habits.

Solution

For the case of twelve grapes, if you have six or less grapes left and it's your turn, you can always win. From there, children expand their thinking, calculating backwards.

Without an Umbrella

Activity

Materials Whiteboard, markers

Set - Up Group work

Wait time One to two minutes

"Samuel went outside without a hat, a coat, or an umbrella, and did not get wet. How could this be?"

For this problem, give students about a minute to think before starting to collect answers. Remind them to doodle quietly or occupy themselves otherwise during this time, even if they are ready. They can try to draw their answer as well.

Teacher's Notes

Write all of the answers on the board, without contesting them or commenting in any way.

When new answers stop coming, ask the students to look at the board. Consider the answers one by one, starting with the ones that contradict the problem statement. Ask the students to consider those answers, and whether they all agree with the proposed solution. If not, write down their explanations and possible contradictions.

Finish with the answers that fit the problem statement. Discuss whether everybody agrees with the suggested solutions.

The focus of this activity is on finding, checking, and comparing multiple solutions.

Student reactions

There will be a lot of discussion. Expect students to forget some parts of the statement as they give answers.

They may also try to negotiate conditions, such as using a cape instead of a coat.

Solution

There may be many good answers. One possible answer is that there was no rain.

Activity **What doesn't belong?**

Materials Print-outs or stickers of objects below

Set - Up Students at their tables, individual work

Wait time One to two minutes

Which object does not belong to the group? Give students about a minute of thinking time before starting to collect answers. Write answers on the board.

Teacher's Notes

Many students initially see only the validity of their own solution, and argue with others. Let them present their solution to the class.

Some students may have difficulty verbalizing the reasoning behind their choice. You may need to gently encourage them. For example, naming parts that they point out (wings, wheels, sails).

Finish by reminding students that a problem may have several good answers. It is our goal to find and consider many answers to each problem.

Students continue to become familiar with multiple solutions, and get exposed to valid points of view that are different from theirs.

Student reactions

Students provide multiple answers, such as:

• The butterfly does not belong, because it is alive.

• The rocket does not belong, because it does not need air to move.

• The sailboat does not belong, because it does not fly.

• The airplane does not belong, because it has wheels.

Activity House Matchstick Puzzle

Materials Craft matchsticks; talking stick (such as toy microphone)

Set - Up Individual work, at tables

Give each student several matchsticks. Let students play with matchsticks for a couple of minutes before you present the problem. In general, children need to play with any new object you introduce. This helps them to explore the properties of the object, and get comfortable with it.

Ask students to build a triangle, a square, a rocket, or a robot, any way they want. Let them compare their designs.

Wait time Five minutes

Move two matchsticks to make the house face the opposite way.

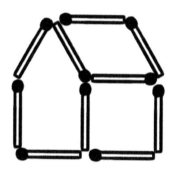

Teacher's Notes

After students are done playing, take the toy microphone and ask for their attention. Introduce the problem. Help each student to reproduce the house; it may be difficult for some of them. Parent volunteers can help. You may also want to use popsicle sticks for younger kids, as they are easier to handle.

If certain students' behavior distracts others, ask these students to move to another table and play quietly. Approach them in a couple of minutes, and introduce the problem one more time.

Eventually, one or two students will figure out the answer. Ask them to wave to you quietly when they are done, so that they will not distract others who are still working on a problem.

In this activity, students exercise their spatial visualization skills, such as mental rotation. Students learn to imagine, build, and deconstruct objects in their minds.

Student reactions

Initially, you may see some confusion about the problem. The students will move fewer or more than two matchsticks. Sometimes, it means that the students are tired, or their attention span is too short. A small snack can help.

Solution

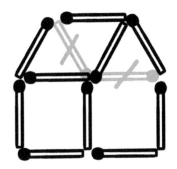

Hands-on Lab **Continue the Pattern**

Materials

Supply of foam animal stickers; paper, pencils

Set - Up

Work at tables, in pairs; patterns can be done in any order

Write the following patterns on the board:

1,1,2,3,5,8, ... ? 2,4,3,6,4,8,5, ... ?

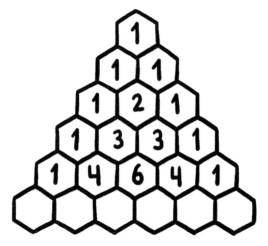

Pascal triangle — add the next line

Ask students to suggest the continuation for each pattern.

Suggest that students come up with their own pattern, and challenge a friend to complete it.

Teacher's Notes

Depending on children's level and attention span, you can cover all or just a few patterns. Encourage children to create their own patterns.

Student reactions

This is a relaxing end-of the class activity. Students can do as much or as little as they want. Many students love the opportunity to create their own patterns, and continue playing with patterns at home. Younger students may get very excited about their patterns, and forget the initial idea they had.

Solution

Dinosaur, flower, dinosaur, flower, dinosaur..

1, 1, 2, 3, 5, 8, 13 (Fibonacci numbers, adding the last two numbers)

2, 4, 3, 6, 4, 8, 5, 10 (Two interlaced sequences, 1, 2, 3... and 4, 6, 8, 10...)

1, 5, 10, 10, 5, 1 (Each number is the two numbers above it added together)

Homework Tangrams

Make animal and spacecraft shapes out of tangram pieces (you may prefer to do this online, through National Library of Virtual Manipulatives, Tangram Puzzles). Do it with friends, too.

Strange statements

Introduction

Goals Provocation; statement analysis; syllogisms; paradoxes; multiple solutions

Materials

Whiteboard, markers; paper, pencils; kids' picture book; poker counters; mini-erasers (for "sand" heap); masking tape; spoon, empty opaque mug; unsharpened pencils or counting sticks

> *"Art is a lie that makes us realize truth."*
> — *Pablo Picasso*

> Mathis' Rule *"It is bad luck to be superstitious."*

Warm - Up # "I am lying!"

Materials Whiteboard, markers

Set - Up Storytelling space (Some circles have special rugs or blankets for storytelling and chats. Others pull chairs in a circle.)

Wait time One minute

Invite kids into your usual storytelling space.

Announce "I am lying!" Keep silent and still for a minute or so, waiting for kids to react and then to stop reacting.

Teacher's Notes

This may be the first time students are introduced to a paradox. You may notice that kids are silent for a while, then start laughing or making jokes. After a while, ask: "Am I telling the truth now?"

Even if kids have a hard time starting, do not provide your own reasons as examples, because this influences their answers too much. The best way to solicit answers is to wait two to three minutes in a relaxed, quiet manner. Kids may ask to repeat the statement several times.

Kids may divide into two camps — those who say "Yes" and those who say "No." Ask the kids from both groups to bring in their objections. Ask each group to defend their position. Alternate between the groups, and record opinions as each person speaks. Ask the kids to look at the board one more time, and take note of their arguments. Promise to discuss this topic next year, to see if they changed their position.

Student reactions

Students will want their own reason in the public list, so record even slight variations.

Solution

This is a paradox, i.e., a self-contradicting statement by design. Do not provide any solution; let it stew.

Activity # Dinosaur Syllogism

Materials Whiteboard, markers; paper, pencils

Set - Up Storytelling arrangement; individual work

Wait time One minute

State the following:

"All dinosaurs have a brain.
I have a brain.
Therefore, I am a dinosaur."

Keep silent and still for a minute or so, waiting for kids to react and then to stop reacting.

Teacher's Notes

This activity introduces students to logical fallacies. The problem is a variation of a classic Syllogistic fallacy.

The kids immediately see the contradiction and start to laugh. Now, you ask them to consider the following:

All kids have brains.
I am a kid.
Therefore, I have brains.

Ask the kids to consider the two statements, and figure out why one works, and the other does not. The kids will suggest some ideas — write them on the board. Provide a pencil and a piece of paper to everybody. Ask them to come up with one syllogism (you do not have to call it such) that makes sense, and one that does not. Parent volunteers may help with writing down the syllogisms. Ask students to read aloud their favorite ones. Suggest that they share them with parents after class.

You may want to share the best syllogisms on your blog or with other math circle organizers (with permission from the parent and child).

Student reactions

Boy says: *"Girls in my family love Lego; I love Lego; therefore, I am a girl,"* and so on.

In some rare cases, children are so used to believing everything that adults say, that they will try to come up with alternative explanations to make sense of the syllogism, such as *"a part of our brain is similar to that of a dinosaur."* An adult may want to address this by saying that math circle is an activity where adults make mistakes, too.

Solution In our language, it is often assumed that *"all"* stands for *"all and only,"* thus leading to confusion. If we would say *"all dinosaurs and only dinosaurs have a brain,"* the statement *"I have a brain and therefore I am a dinosaur"* would be valid.

Activity # Mysterious Book

Materials Whiteboard, markers; kids' picture book; paper, pencils

Set - Up Storytelling arrangement; work in groups

Wait time One minute

Show the kids a picture book. Let them take a careful look through it. Now tell them: *"When nobody looks at this book, it is written in Morse code."* Or, you can use any other language you like instead. Ask them whether there is a way to prove or disprove this statement.

Keep silent and still for a minute.

Teacher's Notes

Ask the students to split into groups according to their opinions about the statement. Let students discuss without interrupting them.

Ask the students to come up with a few examples or draw diagrams of statements that can or cannot be proved, and write them on the board.

Take pictures of the statements and diagrams — you can later share them with students and parents on your class collage, newsletter, or blog.

For younger kids, you can modify this problem by putting a toy animal into an opaque box, and stating *"when in the box, the toy turns into a car."* Ask them how to prove or disprove the statement. Then, ask the kids: *"Make your own strange statement."*

This problem comes as a surprise, since it contradicts pre-existing knowledge, and can not be proved or disproved. This situation often arises in emerging fields of science, when a new phenomenon is discovered and requires existing theories to be entirely reworked. We use this problem to initiate a discussion about which statements can or cannot be proved.

Student responses

Students take some time playing with the book, inventing stories. Let them play for a couple of minutes.

Students may be confused about which statements can and cannot be proved — if so, provide them with more examples, and ask them to sort them into the ones that can and cannot be proved, such as:

"All cats are black."

"Red Riding Hood loves playing computer games."

"Kids in this class enjoy solving problems."

"If I were a robot, I would love fixing cars."

"Five plus five is ten."

Solution

The students can argue forever, since the statement can not be proved or disproved.

Activity **Three Apples Puzzle**

Materials Whiteboard, markers; counters to serve as apples

Set - Up Storytelling space; work in pairs

Wait time Three minutes

Tell the class: *"Two dads and two sons found three apples. Each person got a whole apple. How could this be?"*

Teacher's Notes

Expect that students will request you to repeat the problem several times, until they fully understand the statement. Praise diverse solutions. Put them on the board.

This activity helps the students to analyze their unstated assumptions about the problem statement and introduces them to objects belonging to different sets at the same time (such as intersections in a Venn diagram).

Student responses

Often, children suggest that the people cut the apples. You may want to reiterate that those were whole apples.

One student suggested that they took an apple seed, and grew a new apple tree, so everybody could get enough apples. Another one asked whether a grown adult can be a son, which lead to the solution of the problem.

Solution

There are only three people: grandfather, father, and son. Father is the son of the grandfather.

Activity # Heap Puzzle

Materials Mini-erasers, enough for all pairs

Set – Up Classroom rug, work in pairs

Wait time None; ongoing exploration

Two kids (substitute the names of two students from your class) decided to make a heap out of mini-erasers. However, they soon got confused. They could not decide how many mini-erasers make a heap. Do ten erasers make a heap? How about five? Four? Is one eraser a heap? When does it all change from a non-heap to a heap? Can you add just one eraser to make a non-heap into a heap?

Teacher's Notes

Present the problem. Request that each pair come up with a definition of the heap. Put the answers on the board. Compare the results.

Ask them, how many rhinos does it take to make a heap? Put the answers on the board.

How many Cheerios does it take to make a heap? Put the answers on the board.

Ask the students to compare the results for rhinos vs. Cheerios.

Ask the students to compare Cheerios that are spread out on the table vs. Cheerios that are piled up.

Student reactions

Kids say sometimes that two is a "pair", not a heap; "heap" is mostly defined as starting somewhere between three and six. The answers vary for rhinos and Cheerios.

Some children do not consider anything that is spread out on a surface to be a heap, but as soon as the very same objects are piled up on top of each other, they call it a heap.

Solution

Because heap is ambiguously defined, each individual must come up with a rubric for what constitutes a heap. This activity shows children the limitations of everyday language, and the need for exact definitions. The dictionary definition for a heap is "a collection of things laid in a body, or thrown together so as to form an elevation."

Activity # Elevator Puzzle

Materials Masking tape, counters

Set - Up On the rug or floor, make a large 10x10 grid with masking tape.

All class works as one group; present the problem as a skit. Let students move the counters "up and down" the grid that represents the building.

Wait time Three minutes

In a ten-story building, only one person lives on the first floor, two people live on the second floor, three people live on the third floor, and so on up to the tenth floor where ten people live. On which floor does the elevator stop the most?

Teacher's Notes

This is a good exercise in process modeling, which is common in many areas, including computer sciences. Depending on the number of the students in your class, you may either ask them to represent the people in the building, or use counters for that. A grid made of masking tape is a good way to represent the floors. Put one child or counter on the "first" floor and so on.

For younger children, start with fewer floors (four or six).

You may be asked if people visit each other. For simplicity, we assume that these people do not know each other and do not use the elevator to visit each other — if they are friends or business partners, the story becomes more complicated. There is also no helicopter landing pad on the roof...

Student reactions

Some argued for the tenth floor, because it has most of the people. Some suggested the fifth floor because it's in between. Finally, some students reasoned that everybody should stop at the first floor.

Solution

The first floor, as everybody uses it to enter and leave the building.

Activity # Apple Tree

Materials Paper, pencils

Set - Up All individual work; students back to their tables

Wait time One minute (Do not take in the answers that are given until the wait time is over.)

Twelve pine cones grew on an apple tree. After a strong wind, ten pine cones fell down. How many pine cones are left? Do not say your answer right away. Draw your solution on your paper, quietly.

Teacher's Notes

Problems require an open-minded analysis – sometimes, what you thought was a problem may turn out to be a joke or a mistake.

Jokes also make kids relax, let go, and generate more diverse solutions. This way, the kids are also gradually getting used to consider the problem as a whole before starting to thoughtlessly crunch numbers.

Student reaction

The kids will be coming up with numbers. Give them time. Ask them to draw the problem. Now let them consider their drawings and tell you whether something is wrong.

Solution It's a joke! Pine cones do not grow on apple trees!

Coffee Spoon Puzzle

Activity

Materials Paper, pencils; spoon, empty opaque mug

Set - Up Storytelling space

Wait time One minute

"Now we are going to play a lateral thinking puzzle. I will tell you something, and you will be taking turns asking me yes or no questions. Ask only questions that allow for yes or no answers. Now, let's start. Listen carefully:

This morning, I dropped my spoon into the coffee (act it out with your spoon and mug) but when I picked it up, it was absolutely dry. How could it be?"

Teacher's Notes

It is a good exercise for the kids to consider different possibilities and exercise divergent thinking. Point out that their difficulty rests in some unstated assumptions they make about the problem.

Student reactions

Students may forget and ask questions that cannot be answered with "yes" or "no." You can gently remind them by answering "yes" or "no."

Solution

The mug was full of freshly ground dry coffee

Activity # What do all people on Earth do simultaneously?

Materials Whiteboard, markers

Set - Up Storytelling space

Wait time One minute

Ask the class, *"What do all people on Earth do simultaneously?"*

Teacher's Notes This is an open ended problem with multiple solutions. Record the answers on the board. Ask the students to provide counter examples for suggested answers. For example, if they say "hear," discuss whether everybody can hear.

Student reactions

If students say there is nothing like that, ask them to close their eyes and think more. Usually the students quickly generate a an amazing list of satisfactory solutions: breathe, feel, heat the air around them, etc.

Solution

One of possible solutions: **Live.**

Hands-on Lab # Can you make this?

Materials Unsharpened color pencils (four pencils per kid or per pair)

Set - Up Students back to their tables; students can work individually or in pairs

Wait time Three minutes

Ask the students if they can make the following:

Teacher's Notes

It is healthy to introduce a certain amount of impossible problems, for the same reason we introduce paradoxes: they put the mind to work and prompt wonder. Do not tell the children the solution. Listen to what they say, and write it down in your journal. Reflect on why kids tend not to believe that there are problems without solutions. Maybe they expect adults to always ask them answerable questions. Maybe they are more open to fantasy solutions. Maybe they have stronger divergent thinking than adults.

Student reactions

Most students do not believe that there are problems without answers. Some students may be captivated by the problem for a long time.

Solution

There is no solution (unless these are rubber pencils)

Homework Optical Illusions and brainteasers

Share your favorite questions with friends and family. Ask your parents what they think makes a heap.

Create your own fun story that can be solved by asking "yes/no" questions and share it.

Explore optical illusions. What do optical illusions have in common with strange statements?

Create your own brainteaser, syllogism fallacy, optical illusion or strange statement, and share with friends.

Tiling puzzles

Introduction

Goals Verbalizing visual-spatial concepts; creative use of materials on hand; exploration and use of newly discovered properties

Materials Tiling patterns by M.C. Escher or others; Pentomino set (jar of seven sets can be bought at Amazon); rhino puzzle print-outs; paper, pencils; scissors; camera; Chinese checkers board/printouts, multicolored counters; tall hat with a brim (like a wizard's hat) or a non-transparent grocery bag; whiteboard, markers; wooden sticks; timer; worksheets; list of debate rules (see Appendix A); awards

> *"Mathematics is the art of giving the same name to different things."*
> — Henri Poincare

> *"We adore chaos because we love to produce order."*
> — M.C. Escher

Escher tilings

Warm-up

Materials Pictures of Escher's tiling patterns and tiling in nature

Set - Up Storytelling set-up; group work

Time Five to seven minutes.

Review the pictures. What happens in the picture? Why is it interesting? Introduce the term "tiling." Where can one find tiling in nature? Try to draw a tiling pattern yourself!

Teacher's Notes

This activity helps to see new structures in familiar situations, and opens up the opportunity for creative work. Emphasize that the group has just found a new way to look at familiar things. Challenge your students to discover other ways for looking at things differently in the future. Take pictures of student drawings and make a collage with them.

Student reactions

Students experience a kind of enchantment looking at an Escher tiling for the first time, due to the cognitive dissonance created as "regular" patterns produce seemingly complex effects. Students may need to see multiple examples of tiling patterns to get started on their own projects.

Solution

Some examples of tiling from the world around us are crystals, cellular structures, scales on reptiles or fish, quilts, ornaments, etc.

Activity

Pentomino Rhino

Materials Pentomino set for each student; "rhino" shape puzzle printed out or traced on the board (below); camera

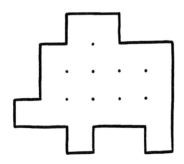

Set - Up Individual work at tables

Time Five to seven minutes

Ask each student to clean a space on the table. Give each student a set of pentominoes. Ask the students to re-create the "rhino" using any four pentomino pieces.

Teacher's Notes

Pentominoes are sometimes a bit too exciting, and open activities excite kids even more. Over-excited kids may start throwing pentominoes or showing their creations to everybody in distracting ways. Starting with a specific shape, rather than letting kids go and do any shape right away focuses and calms their minds. Pentomino puzzles are a variant of tiling puzzles. Arranging pieces to solve the puzzles develops analytical and combinatorial skills.

Student reactions

Not all kids can successfully re-create the shape, as there are so many pieces in the set. This is a good opportunity;

you can invite them to keep playing with pentominoes at home.

If the activity is too challenging, ask the child to close their eyes and sit quietly for a minute.

Some students stop after finding a solution. Encourage them to find more solutions, and share with the class.

Ask students to compare their solutions. Take pictures of each solution.

Solution

There are multiple solutions. For example,

Activity **Pentomino City**

Materials Pentomino set for each student; camera

Set - Up Individual work at tables

Time Five to seven minutes

Tell the students to try and make a house with windows out of the pentominoes.

Teacher's Notes

This activity is easier than the previous one. Free play helps to develop intuition about new objects and their properties. You may want to take pictures of students' creations, or ask a parent volunteer to help.

Student reactions

Kids love creating shapes out of pentominoes and often want to share their creations with friends. Be prepared for active, excited conversations!

Activity # Guess my shape!

Materials Pentomino set for each student

Set - Up Working in pairs on the rug

Wait time Seven to ten minutes

Arrange students in pairs. Put each pair back-to-back on the rug, and ask one of the players to build a figure out of all or some of the pentominoes, without showing it to another player. Next, ask the children to take turns explaining to the other what she has built. The other player tries to re-create it, without peeking. Mention the honor code: no cheating or peeking! Children can pretend that they are explaining their designs on the phone.

Teacher's Notes

This is a geometry version of the "telephone" game. Describing the shape without looking at it helps students learn to verbalize spatial details, invent new words for shapes, and cooperate with partners. It also highlights the importance of establishing clear and consistent terminology in math and science. For kids who may have difficulty with this activity, limit the number of pentominoes to three or four.

Student reactions

This particular challenge is a very engaging activity, so children are likely to ask for several rounds. Bring their attention to the difficulty of finding the right descriptions for the placement of the shapes, as well as for unusual shapes. Discuss the parallels between the work of an artist and of a mathematician.

Hands-on Lab **Small paper, giant hole?**

Materials Paper; scissors

Set - Up Individual work at tables

Wait time Five to ten minutes

In this activity, children will be experimenting with paper, scissors, and the third dimension!

The task is to squeeze yourself through a hole in a piece of ordinary printer paper.

Teacher's Notes

Give children ample time to play. Help children implement their ideas. Ask them to share anything interesting they discover about paper, and put up interesting samples for everybody to see. Write down anything they notice about holes, folds, cuts, and lines. Assure students that there is a solution.

Tell children that the part of math dealing with exotic and weird holes, loops, and lines is called topology.

Practice yourself before showing the students how you do it. Show the solution to the students. Ask them to duplicate it.

Student reactions

After several trials, most students will complain that it is impossible. Let them explore (and complain) for about five to seven minutes.

Young students may lack the manual dexterity to cut the paper properly, so a parent volunteer will be helpful. Encourage children to share this trick with their friends.

Solution

There are many possible ways of solving the problem. All involve transforming the paper into a giant loop (see below).

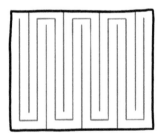

Activity Chinese checkers

Materials Chinese checkers boards/print-outs and play pieces (multicolored counters)

Set - Up Work in groups (two to four people)

Time Five to seven minutes

Chinese checkers are played on tessellated board.
Explore two types of boards: regular and pentagonal
(below)

Which one is easier to play on? Why? Develop your own
play boards, and try to play Chinese checkers on them.

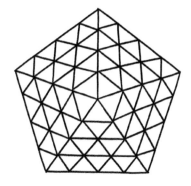

Teacher's Notes

This activity encourages students to explore and
compare, and introduces them into the world of board
game design. It also provides a relaxing end to the class.

Take pictures of the students' boards and ask them to
vote for the best board. Write down their explanations
on the whiteboard.

Student reactions

Not all children know how to play regular Chinese
checkers. Invite children who are new to the game to
join your group so that you can teach them. The boards
may not be just any orthodox grid, but instead must be
composed of irregular semi-patterns. These grids tend
to look organic, like scattered leaves. Children may start

to make up their own rules of play as well as boards. This activity may inspire children to continue designing more games at home.

Homework More Tiling fun

Play Cathedral or Blokus board games. Construct your own pentomino puzzles and share with friends.

Problem debate

Introduction

Goals Elaboration; idea defense; team work

Materials

Tall hat with a brim (like a wizard's hat) or a non-transparent grocery bag; whiteboard, markers; paper, pencils; counters, safe matchsticks or wooden craft sticks; timer; list of debate rules (see Appendix A); prizes

Problem debate is great around the holidays. We host this activity on Halloween. The kids come in costume, and relatives are welcome to observe the class. It is also a great time to take pictures and videos of the class for your website page.

If possible, allow an extra half hour for this class. Remind parents that children should be on time; if not, they may not be able to join a team.

> *"Use soft words and hard arguments."*
> (Proverb)

Warm - Up **Choosing Teams**

Materials Whiteboard, markers; strips of paper to write team names on; hat or bag

Set - Up Help children form teams by arranging chairs around the tables ahead of time, in groups of three

Wait time Two to three minutes

Explain to the students what to expect during the lesson:

"Today we are going to split into teams. You will choose and debate some problems, first within the teams and then between the teams. Decide how you want to solve the problems. You can work together as a group, or individually at first, but eventually the whole team needs to agree on the solution. Now, you will need to find two partners, and name your teams, so we can track scores by team names."

Teacher's Notes

When all teams are chosen, ask them to sit in different areas, so that teams do not disturb one another.

Draw team names from the hat and write them on the board to determine the order in which they will present.

Student reactions

Students often choose their favorite animals, movie and game characters, or fancy math and science terms to name teams. If different team members like different words, they can combine them into one long name, which children find funny.

Activity **Chairs at the Wall**

Materials Worksheet, counters

Children can use counters such as poker chips to place identical chairs in a rectangular room on their worksheet.

Place four chairs along the walls of a rectangular room, so that there is an equal number of chairs touching each wall (one point per solution).

Place eight chairs, so that there is an equal number of chairs touching each wall (two points per solution).

Place six chairs, so that there is an equal number of chairs touching each wall (three points per solution).

Place ten chairs, so that there is an equal number of chairs touching each wall (four points per solution).

Teacher's Notes

These problems provide exercise in several areas: logical thinking, spatial thinking, and analysis of unstated biases. Students learn to recognize their unstated assumptions. For example, they frequently believe that chairs can be only in the middle of walls, rather than corners. The groups can analyze the difference made by putting chairs in corners.

These problems are time-consuming, and children are eager to debate them.

Student reactions

It is hard for some students to realize that chairs can be in the corners.

Students find it interesting that you can use either walls or corners with four and eight chairs. If the presenting team misses a solution, encourage other teams to find other solutions.

For the problem with ten chairs, mixing corners and walls seems confusing. Students may argue that because there are four walls, there must be twelve chairs — even after seeing the solution. Believing is seeing. You may want to discuss this psychological phenomenon with your students.

Solution

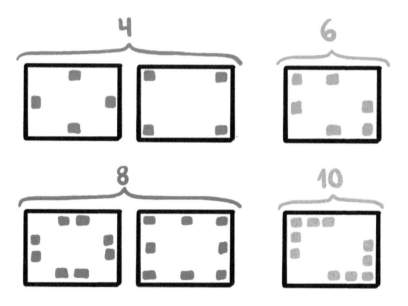

Activity **True expression**

Give out two points for each correct solution.

Use + and – to make a true expression:

4 3 2 1 = 8

Teacher's Notes

Many kids love this problem, as they exercise their pattern-finding skills. Working this problem produces the same kind of pleasure as arranging mosaic pieces into a design.

As a population, girls tend to be more adept at precise calculations than boys, but boys are more motivated to compete. Between these two effects, you may observe some interesting gender differences; just make sure not to discuss them with the kids since this may affect their performance.

Student reactions

Like Sudoku, the correct solution to this problem takes a while to find among many possibilities. However, it is very easy for kids to check for themselves if each conjecture is right or wrong. Such self-correcting problems support independent, quiet work that is meditative for students and relaxing for teachers. If kids want to use parentheses or other operations, they can try that too.

Solution

$4+3+2-1=8$

Activity   Which is Correct?

Award one point for each correct answer.

Ask the class: *"Which is correct: 'seven and five IS thirteen' or 'seven and five ARE thirteen?'"*

Teacher's Notes

The purpose of this problem, and similar jokes, is to make students aware of their approach to reading and critically analyzing problem statements. Students become aware that an innocent statement may turn tricky. This awareness comes in handy when more advanced problems present division by zero, extraneous roots, and other hidden dangers.

The best approach for the opponent team in this case is to present a counterexample (which can also serve as a solution).

Student reactions

Students typically start arguing about the grammar.

Solution

This is a joke;

7+5=12, not 13

Activity   Cavity Conundrum

Award two points for each correct answer.

A boy went to the dentist to have his cavity filled. The boy was the dentist's son, but the dentist was not his father. How can this be?

Teacher's Notes

The purpose of this problem, once again, is to help students recognize and analyze their unstated assumptions.

Student reactions

Students start thinking along the lines of "adopted child." Sometimes they assume the dentist or the son did not know the truth.

Solution

The dentist is the boy's mom.

Activity **How many triangles?**

Award two points for each correct answer.

Ask the students how many triangles they see:

Teacher's Notes

Most of the time, children will present a wrong answer. Write down the answers from different teams on the board. Request the team with the smallest answer to present first, so that other teams can point out more and contribute.

Student reactions

Students provide numbers ranging from six to sixteen. Some kids count only small triangles, but not larger triangles made out of multiple small ones. It is common to have several teams with several different answers, all of them wrong.

Solution

Thirteen triangles

Activity ## Mr. and Mrs. Boo

Two points will be given for each correct answer.

Ask the students: *"Mr. and Mrs. Boo have two sons. Each of the boys has a sister. How many kids are there in the Boo family?"*

Teacher's Notes

This problem provides a great example of graphic analysis. Instead of correcting the children, ask them to draw their solution on the board, indicating boys and girls in the family. You may suggest that they use two types of colored arrows, to indicate the different relationships. Ask the children to contemplate the picture.

Note: this problem is a development of the previously discussed problem from "Introduction to thinking" lesson, "dividing apples between two fathers and two sons." If you have time, ask the students whether they see a connection.

Student reactions

Most of the kids say there are four kids in the Boo family. Help them to see the contradiction.

This problem takes kids into the world of more formal logic, where the meaning of "each of them" is different from the daily meaning of the phrase. Acknowledge that thinking of four kids does follow the daily, everyday use of language. But invite children to contrast this to stranger, and possibly funnier language of mathematics.

Solution

Three kids; two boys and one girl

Activity **Matches and Triangles**

Materials Safe matchsticks/craft sticks

Give three points for each correct answer.

Ask the students to move three matches to another location to get five triangles.

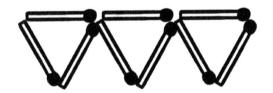

Teacher's Notes

Give ample time for this problem. Ask the children to re-read the problem statement, and discuss their assumptions, pointing out the problematic ones.

This problem is very similar to the "How many triangles?" activity. Ask the students how the two problems connect when you discuss the solutions.

Explain that problems connect to one another in a giant and intricate web. The more problems you solve, the more of the web you get to see.

Student reactions

You get a wide range of responses to this problem. Sketch them, or take pictures.

Kids usually come up with four identical triangles, and then stop.

The challenge here is to understand that the triangles may be of different sizes, and they may be inside each other — a counterintuitive idea for many students. This problem is related to the problem of counting triangles in a previous activity.

Solution

Equilateral triangle with a side length of two matchsticks (four small triangles, and a big one).

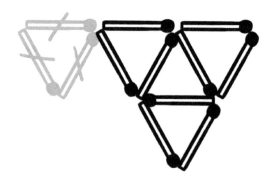

Contest Finale

Materials

Prizes (dice of different shapes, Moebius strips, or slide puzzles work well)

Set - Up

Classroom rug

Ask all the teams to come to the classroom rug. Invite everybody to shake hands.

Announce the team scores. Congratulate each team, and say something special about each one. Distribute the prizes.

Homework

Students may want to work on the problems that they did not have a chance to discuss in class.

Infinity

Introduction

Goals In-depth focus on exploring a single topic; recursion; cycles

Materials

Clipboards, pencils; toy microphone or other "talking stick"; paper, pencils; small mirrors with dull edges; pictures of infinity (see "Drawing Infinity" below);

pictures of fractals; dice; colored pencils; masking tape or chalk.

Infinity is one of those topics engaging for people "from age three to ninety-nine" since it relates to deep — yes, infinitely deep — concepts.

"What defines a successful discussion? It is when you walk out with more questions, than answers"
— Peter Senge

"The infinite is always unruly unless it is properly treated"
— James Newman

Warm - Up "What is Infinity?"

Materials

Toy microphone or other "talking stick"; whiteboard, markers; clipboards; paper, pencils

Set - Up

Classroom rug or storytelling place

Wait time

One minute. Don't give the microphone to anyone for about a minute after you pose each question. Remind children that this is quiet thinking time. Invite them to doodle or record answers if they lose patience.

Ask the students:

"What is infinity? Explain."

Teacher's Notes

Welcome each and every response, and write it on the board. The fact there are many different answers invites an open-ended discussion. It is not easy to come to a definition of infinity. The word itself comes from a Latin word meaning "no boundaries."

If many students can't get going after two minutes or so of quiet thinking, ask related questions to prompt discussion. Does infinity exist in reality? If yes, where can we find it? If no, why not and where does it exist?

Student reactions

Children may discuss whether infinity is a number at all. This is a very promising discussion, because children need to be learning many math entities beyond numbers and operations. Most of the time we use infinity as a concept. In mathematics, infinity can sometimes be treated as a number, without ever becoming a true number. For example, mathematicians can talk about operations on infinity, such as, "What is infinity plus five?" This leads to paradoxes children love (below).

Some children refer to infinity as the biggest number they know, such as googolplex. If this happens, ask them if it is possible to add one to googolplex. If they agree that it is possible, ask them "what is the biggest number?" again. They should be able to reason that now googolplex+1 is the biggest number, and that you can keep adding one. Some will make the claim that numbers go on forever. This is a difficult concept for a child.

Drawing Infinity

Activity

Materials Paper, pencils

Set – Up At tables, individual work

Wait time Three to five minutes

Suggest that students go to their tables, and draw something that is infinite. If many children seem stuck after a minute or so, suggest they draw something that goes on and on forever, or has no boundaries.

Teacher's Notes

Give students three to five minutes to draw, then come to the rug. Ask them to sit in a circle and take turns sharing and explaining their work. Welcome friendly comments from other students.

Be a part of the circle as well. When it is your turn to share — don't be the first or the last — show three or four prepared images or animations from science, modern and ancient cultures, and mythology.

Suggestion for your pictures: oroboros, the scale of the universe animation, the recycle symbol, the walk cycle animations, mandalas, and fractals.

Student reactions

Students come up with wonderful oscillating lines, repeating patterns and simple fractals.

Some students write down the symbol ∞ as a way to "draw infinity." Congratulate them on knowing the symbol, and explain the difference between symbols and drawings. A useful analogy is the difference between a student's name (a word), and the student's portrait (a drawing). Ask students, now that they wrote down the "name" of infinity, to draw its "portrait."

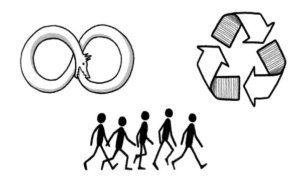

Activity **Recursive Poems**

Set - Up At tables, individual work

Sing or repeat:

"I know a song that gets on everybody's nerves, everybody's nerves, yes, everybody's nerves. I know a song that gets on everybody's nerves, and this is how it goes!..."

Repeat the song until it gets annoying! At this point, ask the students what they think.

Teacher's Notes

After the discussion, ask students if they know other never-ending songs, repeating scenes in movies, or repeating patterns in art, music, or dance.

Student reactions

Because of previous discussions, students will talk in terms of infinity, for example:

This song doesn't go on forever; it ends when the person who is singing it dies.

Activity # Mirror Images

Materials Small, safe mirrors (two mirrors per pair of students)

Set – Up In pairs; work at tables

Wait time Three to five minutes

Ask two children volunteers to hold two mirrors facing one another. Put a small object, such as a toy animal, between the mirrors. Ask the students to count the number of reflections of the object.

Teacher's Notes

Allow three to five minutes for this activity. Let each kid hold the mirrors at least once, because they often want to experiment with angles.

Students reactions

Students find this "homemade infinity" fascinating. They all want to volunteer as mirror holders, and are amazed at the number of reflections.

Activity # Fractals

Materials Whiteboard, markers; paper and colored pencils/crayons; fractal images print-outs (you can get a lot of fractal images online)

Set – Up Individual work at tables

Wait time Two to three minutes.

Introduce students to self-repeating patterns, objects or processes, such as live cell replication, nesting dolls, etc. Tell them that a self-repeating pattern is called a fractal, and the iterations (repetitions) can go forever.

Share some images of fractals.

Teacher's Notes

Invite the students to make fractals of their own, and color them in different ways. Remind them to be creative with their shapes.

Ask the students: *"What do fractals have to do with infinity?"*

Student reactions

Fractals are mesmerizing and easy to produce. Students enjoy drawing and observing them. While students often have a good intuition about the relationship between fractals and infinity, they may not be able to verbalize it. Instead, invite them to point it out in their pictures.

Activity **Squares and Rectangles**

Materials Whiteboard, markers; paper and colored pencils/crayons

Set - Up Individual work at tables

Wait time One to two minutes

Tell students to get ready for a question. Ask: *"Are there more squares than rectangles?"*

Wait at least a minute before accepting any answers. Remind students to take notes or doodle quietly, even if they are ready, so that others can think.

Teacher's Notes

It is not easy for our mind to compare two infinite sets. While the students may argue that there are more rectangles, as squares are a type of rectangle, there is an infinite number of both.

Student reactions

Some students will argue there are more squares than rectangles, and some that there are more rectangles than squares. For some reason, young students see squares as much more comforting and familiar than rectangles. The students who are more mathematically mature may argue that a square is just a type of a rectangle, so squares are just a subset of the rectangle set, and so there are more rectangles than squares. It is very important to acknowledge this logical conclusion, and to ask the children to share their thoughts with the class.

This line of thinking sounds very convincing to the rest of the class. After you let the children convince the class, ask them to compare two finite sets — e.g., a set of three objects and a set of five objects. They will find the one-to-one correspondence, and point out that one set is bigger. Now you can return to the original question, and ask them how could they go about finding a one-to-one correspondence of sets?

Solution

There is an infinite number of squares and rectangles. You cannot compare infinities as you do numbers. But do not despair — there are secret ways of comparing infinities, which were invented by mathematicians!

Infinite and Unlimited

Activity

Materials Classroom rug with a grid made of masking tape

Set - Up Classroom rug; group discussion

Wait time One minute

Ask the students:

"Should something infinite also be unlimited?"

Teacher's Notes

After the students have come to an agreement, ask for a volunteer. Ask the volunteer to make a really big step (use the rug with the grid to better measure the step length). Now ask the volunteer to make repeatedly smaller steps, with each step half as small as the previous one. Ask the students whether the volunteer will be able to cover an infinite amount of space with an infinite number of steps.

This illustration of repeatedly splitting a rectangle in half may also be helpful (Alternatively, you may want to tear apart a piece of paper).

This serves as a basis for mathematical anecdote about catching a lion in the desert. To catch a lion in the desert, split the desert in half. The lion is either on the left or on the right of your line. Let's assume that it is on the right. Now split the right half in two. The lion is now in one of the quarters. Let's assume that it is in the bottom right quarter. Continue until you have corralled the lion into a sufficiently small enclosure.

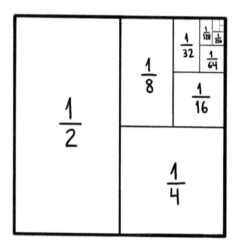

Student reactions

Most of the time, students reply that infinite is also unlimited. You may want to ask for specific examples.

Activity # Hotel Infinity

Materials Whiteboard, markers

Set - Up Storytelling place, group discussion

Wait time One to two minutes

This is an example of Hilbert's Paradox. Tell the students the following story:

"A long time ago, in a galaxy far away, there was a hotel with infinitely many rooms. Fred, the owner, ran his business really well; every room was occupied with a guest! There were all sorts of aliens

in different rooms. You can draw them while I tell the story. Then, one day, another alien arrived and asked for a room. All the rooms were taken, but Fred did not want to turn a guest away. What do you think he did?

Oh, Fred did something very clever! He took his speaker, which broadcast to every room in the hotel, and said: "Attention everybody! Please move to the room with the next number!" So, the guest from room 1 moved to room 2. The guest from room 2 moved to room 3. The guest from room 3 moved to room ..., and the guest from room one million moved to room ... But what happened to room 1?"

Teacher's Notes

This story sounds like a paradox, and paradoxes make our brains work. The surprise of the solution helps to keep the motivation alive. Children love to discuss this problem, exercising their divergent thinking abilities. Leave ample time for the discussion. If children are frustrated, ask them to act out the aliens.

If the children are not tired, you may want to continue the discussion with the "two more guests" extension problem: *"Everything was quiet at Fred's Hotel Infinity, with infinitely many rooms, and infinitely many aliens. But guess what happened at the hotel the next day? Two aliens arrived and asked Fred for rooms! What do you think Fred did?"*

Student reactions

Students suggest reconstructing the rooms (e.g., split a room in half), or claim that Fred would have to get rid of someone. They also come up with all types of fantasy solutions, such as that aliens have an ability to dissipate, turn invisible, or change sizes.

Do not forget to write down their answers – they are fun to share with parents and peer teachers!

Hands-on Lab **Dice Roll**

Materials Dice; chalk or masking tape

Set - Up

Group work; if the weather is good, take the children outside and draw a line with chalk; otherwise, take them into the hall and make a line with steps counting from using a masking tape.

Ask for a volunteer to walk the amount of steps shown on a die. Ask for volunteers to roll the die, to read the die number aloud, and to count the total number of the die rolls.

Write down everybody's names and predictions.

Wait time Two to three minutes

Explain the problem to the students: "You have a number line with twenty steps extending in each direction from zero. Zero is your starting point. Your goal is to get to the number 20 on the line. You throw a die to get the number of steps for each turn. Every time you throw the die, you change the direction of your movement to the opposite one (it is OK to walk past zero to the left, too). Will it be possible for you to reach the number 20 on the number line? If yes, how many die rolls will it take?"

Teacher's Notes

This is an experimental problem that allows children to move around, to take turns with the manipulatives, and to develop their mathematical intuition about randomness, as well as get a taste of negative numbers.

Depending on the level of the students' arithmetic skills, or if the children seem to get tired, you can change the number of steps to ten.

Student reactions

It is fun for the children to see how they differ in their predictions. They become big fans of their own predictions, and hold to them, as they are watching the game.

Sometimes students keep yelling their predictions repeatedly, rather than experimenting or sharing the logic behind the predictions. Remind them that answers get more and more correct when you repeat them many times, as loudly as you can!

Homework Doodling and fractals

Watch beautiful "math doodling" videos by Vi Hart on YouTube and doodle along.

Explore fractal images online, such as PBS NOVA fractals.

Symmetry

Introduction

Goals Symmetry in life, science, and the arts; our brains' preference for symmetry

Materials

Demo materials transparency, dry-erase or permanent marker; tack board, tacks

Group materials paper, color pencils or crayons, erasers; camera; paper or coffee filters; safe scissors; small mirrors

Symmetry has a special appeal to us as humans. We tend to notice it and to adore it. Why are we inclined to pay special attention to it? One hypothesis comes from information processing theory.

> *"The mathematical sciences particularly exhibit order, symmetry, and limitation; and these are the greatest forms of the beautiful."*
> — Aristotle

Warm - Up "**What is Symmetry?**"

Materials Paper, color pencils or crayons

Set - Up Individual work at tables

Wait time Two to three minutes

Greet the students, and ask them to take their seats and draw a symmetrical object.

After the students draw something, ask them to show the symmetry. For example, they can fold the paper along the symmetry line, or trace the reflected and rotated parts of the drawing with different colors.

Start the discussion with: ***"Today, we are going to discuss symmetry. What is symmetry?"***

Teacher's Notes

Write down the answers on the board — with examples, if appropriate.

In the middle of the discussion, ask: "The word "symmetry" sounds similar to "same-metry" — do you think this is just a coincidence? Why or why not?"

If during the discussion the students do not mention palindromes, you may want to introduce a couple of examples yourself, to point out the existence of symmetry in other fields.

Student reactions

Symmetry is:

• Fair sharing between two friends.

• A word that reads the same both ways (palindromes, such as "radar" or "level").

• When you fold it, it meets together.

Activity
Types of Symmetry

Materials Transparency, markers; tack board and pins

Set - Up Individual work at tables

Wait time Two to three minutes

Show and discuss different types of symmetry (reflection, rotation, and point symmetry). Invite students to look at their drawings and sort them in piles according to these attributes.

Teacher's Notes

Review the final piles. Ask the students to consider which pile has more student drawings, and why.

Take a rotational symmetry drawing that is more or less correct. Place it on your board, cover with a transparent sheet, and pin the two sheets through the center of rotation. On the transparency, trace the major parts of the drawing with a dry erase or permanent marker. Then rotate the transparency to show rotational symmetry. Be careful with pins around children! You may also want to demonstrate point symmetry the same way.

Invite students to draw a fantasy world where everything is symmetrical, and then a fantasy world where nothing is symmetrical. They can share their drawings with each other to catch mistakes. Ask which picture looks more natural, and which picture was more natural to draw.

Student reactions

Most of the time, students of this age only know reflection or "mirror" symmetry. However, they love other types of symmetry if given a chance to explore.

Discussion Symmetry in Faces

Materials Paper, pencils; camera

Set - Up Individual work at tables

Wait time Two to three minutes

Ask students to draw a symmetrical face and an asymmetrical face. Which one is more pleasant? Which one is more interesting?

Teacher's Notes

Take pictures of students' creations. Laugh. Discuss how people find familiar things soothing. Discuss how people find strange things — which provide new information — more interesting. Ask them to vote on the most attractive, most ugly, and most interesting faces.

Symmetry allows us to save mental resources in learning, recalling, and operating information. For example, we really need to remember only half of a face, because the other half is almost the same. However, most objects — including faces — are not perfectly symmetrical. We have to be careful not to simplify reality too much.

You can share the following with the students: If you take a photograph of a person's face and split it down the middle, you can make an entirely different face by reflecting each half.

Student reactions

Kids have a lot of fun making the asymmetric face as weird as possible. Many will say the symmetric faces are more attractive, but asymmetric more interesting.

Discussion

Noticing Symmetry

Materials

Whiteboard, markers

Set - Up

Storytelling place such as a classroom rug

Wait time

One to two minutes

Ask students to find some symmetrical objects around the room. This is usually an easy task, and they will find many examples. Ask: **"Why do we pay so much attention to symmetry? What does symmetry mean to you?"**

Teacher's Notes

Ask students if they think animals notice symmetry, too. Record student ideas on the whiteboard. The goal of this task is to help students make symmetry personally relevant.

After listening to some answers, share that symmetry perception has been demonstrated in humans, birds, dolphins, and apes. Even insects, such as bees, favor symmetrical over asymmetrical patterns. Humans view symmetry as beautiful, comforting, and informative. We associate symmetry with feelings of peace and harmony. In science and math,we often turn to symmetrical models because they simplify the world for our analysis.

Student reactions

Students are eager to find symmetrical objects. They may have profound ideas about the "symmetry sense" in humans and animals.

Discussion # Tyger, Tyger!

Materials Pictures of tigers

Set - Up Group Activity

Wait time One minute

Read some of William Blake's "The Tyger" to students.
The first and the last stanzas are:

> *Tyger! Tyger! burning bright*
> *In the forests of the night,*
> *...*
> *What immortal hand or eye*
> *Dare frame thy fearful symmetry?*

Show a picture of a tiger head on, and a tiger in profile,
and ask the children: "Which tiger is about to eat you?"

Teacher's Notes

Write down the students' thoughts about each picture.
Follow with a discussion why recognizing symmetry may
have been vital for an early hominid's survival.

Student reactions

Most students will consider the tiger in the first picture more dangerous, because that tiger definitely pays attention to you. The second tiger may ignore you and follow its path.

Activity **Continue the Series**

Materials Pattern worksheet, pencils

Set - Up Individual work at tables

Wait time Five to six minutes

Ask the students to continue the following series:

Teacher's Notes

Younger kids may lack the manual dexterity to draw the correct figure even if they know the answer – allow extra time for this activity.

Take time to consider all of the suggestions and gently offer counterexamples. If they are sure that they have the right answer, ask them to continue the pattern even further. Suggest not looking at the objects as geometrical figures, but rather solving the problem as a cryptographic puzzle.

It is not easy for the students to recognize the pattern (mirror symmetry) in this problem. The topic of the lesson — symmetry — may be used as a hint.

Student reactions

Most of the time, kids try come up with their own patterns to continue the row, many of which are very ingenious.

Solution

The sequence is made of Arabic numerals and their symmetrical mirror reflection. The next figure is the number five and its reflection

Activity **Build a Triangle**

Materials Paper, pencils

Set - Up Individual work at tables

Wait time Three to five minutes

Draw a triangle such that all 4 dots belong to its sides.

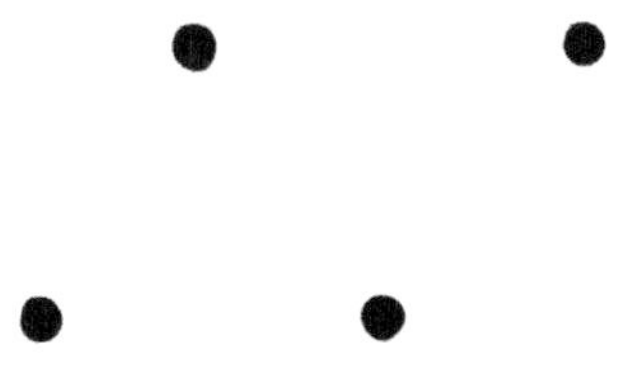

Teacher's Notes

To get the students started, ask them to draw various triangles and non-triangles. Ask them how they know which are triangles, which helps them verbalize geometric properties.

In many "insight" problems we assume symmetry where it is not necessary. The near-symmetry of the points' locations is misleading; it makes us think that all points are vertices.

Student reactions

At first, students see the problem as impossible. Many of them see that the points can be vertices of a diamond (rhombus), and this locks their mind on that particular shape. To get unstuck, students may play with the arrangement of the points.

Solution

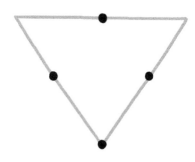

Activity **Mirror Writing**

Materials Safe mirrors

Set - Up Individual work at tables

Wait time One minute

Tell the students that you are going to play a mirror game. Distribute the mirrors to the students, let them play with them for a minute, then ask them to read the secret message below using the mirrors:

Hi, great students of my exciting class!

Invite students to write their own "mirror code" secret messages for others in the group.

Teacher's Notes

Do not point out spelling mistakes! However, do help children notice and correct geometric mistakes (incorrect reflections) by checking with the mirror.

Mention the other name of line symmetry — "mirror symmetry."

Student reactions

Students often start by writing their own names or the names of their friends. Children often make more spelling or computation mistakes than usual when exploring new ideas. Many students can figure out how to read the message without the mirror. This means they visualize or imagine the reflection.

Hands-on Lab # Snowflakes

Materials Scratch paper or coffee filters; safe scissors for each student

Set - Up Individual work at tables.

Show students how to make paper snowflakes by folding the paper and cutting it. Then, invite them to design their own.

Teacher's Notes

Ask parent volunteers to help the kids to cut out their designs. You may want to have a discussion on the snowflakes' types of symmetry with individual students. Consider drawing the center and axes of symmetry on the board.

Student reactions

Students cannot wait to see what their snowflake will look like when they unfold it. Many of them come home and continue making new designs for weeks on end.

Homework Symmetry Artist

Create symmetrical art by hand or using online tools. Look to Escher, Dali, and other famous artists of optical illusions for inspiration. Create ambigrams — words or images that can be read from multiple directions, such as "Math magic" below

Lateral thinking

Introduction

Goals Working backwards; divergent thinking; problem-solving heuristics

Materials

Paper, pencil, eraser; ruler; maze print-outs; "drop that block" handout; masking tape; counters; graph paper or tic-tac-toe grids

> *"I worry much more about unquestioned answers than about unanswered questions."* — Unknown

> *"It's a poor sort of memory that only works backwards."*
> — Lewis Carroll

Warm - Up

Materials Whiteboard, markers

Set - Up Storytelling space

Wait time Two minutes

Read the problem, without stressing the key parts with your voice: *"Mike had eight Lego figures. In one unfortunate week, he lost all of them but five. How many does he have now?"*

Teacher's Notes

This problem is a bit trickier than it seems. The wording misleads students to the automatic reaction of subtracting five from eight, which is incorrect. This problem reminds the kids that they are back to the strange world of the math circle, and teaches them to analyze their problem statements thoughtfully.

Student reactions

The class divides into two groups — those who say "three" and those who say "five." Let the children argue for a couple of minutes, until they figure out the answer for themselves.

Solution

Five

Activity **Answer Guessing**

Materials Paper, pencils

Set - Up Individual work at tables

Wait time One minute

Turn away from the board. Ask a volunteer to write a number on the board. Tell the students: *"I am going to guess your answer without knowing your number"*

Ask the volunteer do the following without reading the results aloud: to add two; subtract four; add ten; subtract the number they chose; subtract five. Say *"you got three as an answer!"* Ask the students how you could possibly know the answer.

Teacher's Notes

This problem introduces children to the problem-solving method called "working backwards." This way of solving problems is counter-intuitive for many kids this age.

Children (as well as most adults!) are not used to undoing steps that were previously finalized. This is an ability that does not come naturally to us, so do not be disappointed if it does not happen right away.

If children are having difficulties remembering the problem, or following the order of the steps, you may want to write the steps on the board. The difficulty, for students, is to "undo" each operation at each step.

You are welcome to use any other combination of numbers or arithmetic operations, to give the students different answers each time. If you have time and the children seem interested, invite them to make a similar puzzle for working backwards. Parent volunteers can help kids write down puzzles and exchange them with their peers.

Student reactions

Some kids use trial-and-error instead of working backwards, trying some number and modifying it if it does not work. Many students solve this problem quickly, as it does not have many steps.

Solution

$x+2-4+10-x-5=3$

Activity **Maze**

Materials

Maze print-outs (you can use the one provided below, or any other of your choice)

Set - Up Individual work at tables

Wait time Four minutes

Solve the maze first from start to finish, then from finish to start. Which way is easier?

Teacher's Notes

Do not distribute the maze worksheets before you are done with the previous activities; it will be distracting.

After the students finish the maze, ask:

"What happens if you switch the start and finish? Does the direction matter?"

Write down the answers on the board. Let the students explain their point of view.

Student reactions

The students provided a set of answers, such as:

• Forward, because I always do it this way.

• Backward, because you remember how you did it.

• Backward, because going forward, you have no clue where you are going.

Activity **Squad of Soldiers, a River-Crossing Problem**

Materials

Masking tape; colored poker chips or counters for each pair of students

Set - Up At tables, working in pairs

Wait time Three to five minutes

A squad of six soldiers must cross a deep river with no bridge. They spot two children playing with a small rowboat by the shore. The boat can only hold two children or one soldier. Can all of the soldiers make it across the river?

Teacher's Notes

Acknowledge the creativity of the students by writing all of the wacky and wild ideas on the board. It is a great example of the divergent, lateral thinking we try to encourage in children. Explain that it will be a greater challenge to focus on the problem as intended: with soldiers getting across the river in that one little boat.

Help children use the available manipulatives to represent each element of the problem (masking tape for a river, poker chips for the characters).

Student reactions

Students who don't see an immediate solution will make many creative attempts to re-state the problem circumstances.

Are soldiers children, and are children soldiers?

Can the soldiers build a larger boat?

Can they just wait for the river to freeze?

When the students say that one of the children can go back, it usually solves the problem, as they figure out the cycle of moves that get the soldiers across the river one at a time, and keep repeating that set of moves.

Solution

Two children cross the river.
One child crosses the river back.
One soldier crosses the river.
One child crosses the river back.
Repeat until all soldiers have crossed.

Activity # Wolf, Goat, and Cabbage

Materials Colored poker chips or counters

Set - Up At tables, work in pairs

Wait time Three to five minutes

A man has to take a wolf, a goat, and a huge head of cabbage across a river. His rowboat has enough room

for the man and one other passenger. While the man is nearby, the wolf and the goat behave. If left alone, the goat will eat the cabbage, and the wolf will eat the goat. How can the man get everything across the river intact?

Teacher's Notes

This is a classic puzzle. There is a alternative version about a fox, a goose, and a sack of corn. Most people are not used to undoing the steps that seem finished. This problem requires taking the same object (the goat) back and forth over the river — a counter-intuitive strategy!

Student reactions

It is not rare for the children to say that it is impossible, at least at the first try. However, given the means to recreate the problem with manipulatives, they quickly find the answer. The older the students, the more difficult it may be for them to deviate from the familiar straight and narrow path to a solution and just play with the problem.

Two Solutions

Move the goat. Move the cabbage. Return with the goat. Move the Wolf. Move the goat.

Move the goat. Move the wolf. Return with the goat. Move the cabbage. Move the goat.

Activity **Brother and Sister**

Materials Whiteboard, markers

Set - Up Group work, storytelling place

Wait time One minute

A brother and a sister come to a river. The river is deep and wide, the current is strong, and there is no bridge. There is a boat that can only hold one child. However, both children crossed the river. How?

Teacher's Notes

This is a good exercise for discovering unstated assumptions. Children usually assume that the kids come to the river from the same side. Support all of the creative ideas, but help children distinguish between making new problems and solving this one. Ask the children to defend their idea in front of the class, with their classmates offering counterexamples.

Student reactions

The students are eager to approach the problem, and tend to generate some wonderfully inventive solutions. For example:

• The sister can cross, then push the boat as hard as she can back across the river to the brother. Reply that this would work, except the river is too wide and the current is too strong, so the boat would float away.

• One of the children can hold onto the back of the boat while the other rows across. Reply that the current is too strong, so it is dangerous.

• Both children get in the boat, but they would alternate jumping really high in the air, so that only one person would physically be in the boat at a time. Reply that it would sink the boat, because of how the momentum works. Also note that this will also prevent them from rowing.

Solution

The two siblings came to the river from two opposite sides! The one with the boat crossed the river, and gave the boat to the other.

Activity **Crossing with no Boat**

Materials "Drop that block" illustration; white board

Set - Up Group work, storytelling place

Wait time One minute

One person came to a river. There was no boat in sight, and he cannot swim. However, he crossed the river. How?

Teacher's Notes

By now, the students have developed assumptions about what constitutes a river. This prevents them from seeing the obvious solutions.

Write down the suggested solutions on the board, without commenting. After the students have exhausted their explanations, offer a few simpler solutions. Discuss with the students how quickly we form the mental habits that prevent us from seeing other solutions.

Share the "drop that block" illustration with them, and ask them to observe their behavior to become aware of their own mental habit formation.

Student reactions might include:

There wasn't a boat in sight, but maybe the person could walk somewhere that was currently out of sight and find a boat.

Build some sort of sling shot using a tree to throw the person across the river.

Walk far enough to go around the river.

Jump from stone to stone, throwing in more so they are close enough until there is a whole walkway.

Solution

Many solutions are possible. There could be a bridge, the river is covered with ice, etc.

"Drop that block" handout

A person below is holding a block of wood. Now, what will happen to the piece of wood when the person lets go of it?		

Fold here; do not let students look at the pictures below, until after discussion

Earth	Underwater	Space
The block of wood will drop DOWN to the ground as it is drawn to earth by gravity.	The block of wood will float UP to the surface of the water because it is less dense than water.	The block of wood will NOT MOVE because there are no overall forces in any direction.

Math Lab **Inverse Tic-Tac-Toe**

Materials Graph paper or tic-tac-toe grids

Set - Up At tables, work in pairs

Wait time Two minutes

Teacher's Notes Ask the students to play one game of regular tic-tac-toe. Now ask them to play a round of inverse tic-tac-toe, where the first player to draw three in a row loses. Ask the children to reflect on the inverse game. Is it easier or more difficult than a regular game? Why?

Discuss how we can gain awareness of our habits, and how to challenge them.

Student reactions

Their moves in inverse tic-tac-toe are slower, and require more thought than in the regular game.

Homework Puzzles

Search for "plank puzzles" online or play the ThinkFun River Crossing game. Design your own mazes.

Festival

Introduction

Goals Teamwork; community building

Materials Dice; print-outs with numbers; game boards, playing pieces for board games; parent bingo print-out; certificate of completion; ribbons or other awards; flyers for next session (if applicable)

> *"It should be noted that the games of children are not games, and must be considered as their most serious actions"*
> — Michel de Montaigne

> *"Learning is all about connections, and through our connections with unique people we are able to gain a true understanding of the world around us"*
> — Peter Senge

Warm - Up

If you've never organized any group events before, the World Cafe design principles may help you to get ready (see Resources).

Greet the students and parents. Invite children to sit around the rug and get comfortable. Explain that you are going to have a problem-solving festival where everybody is encouraged to play and think.

Activity **Dice Magic**

Materials Dice; parent bingo for parents (see Appendix C)

Set - Up Two to three students at a table

Wait time Two minutes (The students may be excited by the parents' presence; remind them of the wait time prior to the activity.)

Tell the students that you have a magic ability to read hidden numbers. Ask them to build a tower out of several six-sided dice. Tell them that by looking at the top face of the top die, you can predict the sum of the dots from all the top and bottom faces in the whole tower, even hidden ones.

Teacher's Notes

The activity invites students to get familiar with dice. This familiarity will come handy later, as they start on probability. Students enjoy predicting the sums of the dots themselves, and are excited about "magic."

Repeat this activity several times, prompting the students to look for some regular pattern. Write students' explanations on the board. Ask others to agree or disagree.

In the meantime, ask the parents to stay at the tables and play Parent Bingo.

Student reactions

Students may argue over who will build the next tower.

Let everybody take turns building.

Solution

The sum of the dots on any opposite faces of a six-sided die is seven and does not depend on how the dice are positioned. Hence, the total on all the top and bottom faces will always be equal to the number of dice in the tower multiplied by seven.

Activity **Friendly Handshakes**

Materials Whiteboard, markers

Set - Up Storytelling space

Wait time Two minutes

Tell the students: *"We will be shaking hands with one another (once with each person) — but wait! Before we shake hands, write down your predictions for these questions. Parents are welcome to make predictions too!"*

A parent volunteer can write everyone's predictions on the board.

If all adults in the class shake hands with one another, how many handshakes will there be?

If all kids in the class shake hands with one another, how many handshakes will there be?

If everybody in the class shakes hands with one another, how many handshakes will there be?

Encourage everybody to shake hands so that they can test their predictions.

Teacher's Notes

This activity may get out of control. To keep it orderly and to help with counting, encourage students to take turns. Alice goes first, shakes hands with everybody, and sits down; Bob goes second and does the same.

It is not easy for the students to come up with an answer. Encourage students to set the lower and upper limit of handshakes possible, as a first step of the solution. Another helpful problem solving technique is to simplify the problem by splitting up into smaller groups of people.

Student reactions

Initial guesses tend to be very far from reality. Let everybody experience this effect several times, because it motivates a sense of wonder. Eventually, they start to notice some aspects of the pattern, such as each person shaking a number of hands that is one less than the total number of people. Do not expect students to provide you with a written formula, but see if they can make correct predictions.

Solution

For a group of ten people, each person shakes hands with nine other people, for the total of: $10 \times 9 = 90$

However, there is only one handshake per two people. The answer is: $90 / 2 = 45$.

The general formula for n people is: $n(n-1)/2$

Children may be able to figure it out, but they are likely to describe it with words rather than symbols.

Activity   # What is Next?

Materials Whiteboard; markers

Set - Up Parents and children split into two separate groups

Wait time Five minutes

Parents and children work as separate groups. Encourage people to discuss the problem within their group only, since it is a competition between parents and students.

Present the following set of numbers to each group. Ask groups to carefully examine the list, and to agree on an answer.

8809 = 6	9312 = 1
7111 = 0	0000 = 4
2172 = 0	2222 = 0
6666 = 4	3333 = 0
1111 = 0	8193 = 3
3213 = 0	8096 = 5
7662 = 2	**2581 = ?**

Teacher's Notes

This is a great exercise in divergent thinking. We are very indoctrinated into seeing numbers as representations for quantities. As a result, we forget that numbers can also be interpreted as shapes.

Student reactions

Students may ask how such large numbers are equal to small ones. Acknowledge that it's an acute observation. The equals sign here works differently than usual. Explain that this often happens in some kinds of mathematics as well as computer sciences. Invite children to figure out what the sign means!

Kids are more observant and less conditioned than adults. They often find the solution faster.

Solution

Count the number of circles in the written representation of each four-digit number.

Hands-on Lab # Board Games

Materials Board games printouts and play pieces

Set - Up "Activity centers" on the tables

Ask students and parents to mix at the tables. Ask parent volunteers to distribute game boards. Invite each group to select a game and start playing.

Some board games you may enjoy:

Ducks in a row, Nim, Star, Swag, Sudoku, Kenken, Connect four, Battleship, Dot-to-dot, Blokus, Chinese checkers, Gobblet, Mastermind

Teacher's Notes

Clarify rules as needed.

This activity fosters parent involvement, and creates more topics for discussion between parents and students.

Student reactions

Students are happy to have this extra playtime with parents. Some of them manage to beat the parents in the games!

Wrap — Up

Materials

Talking stick (such as toy microphone); certificates of completion; ribbons or other awards; flyers for the next session

Set - Up Students and parents at the tables, facing the teacher

Teacher's Notes

Thank all of the students and parents.

Distribute the flyers for upcoming sessions and remind parents about your next courses, if any.

Ask parents and students to share their reflections about the session, starting with the children. Do not force anybody to talk. You can model by recalling funny mistakes (including those of your own), and saying what you liked best about the session. Ask them to name some of their favorite activities from the session. Invite them to share their thoughts on how they have benefited from the activities.

Distribute certificates. You may also present each of them with an award, such as a ribbon. Do not use candy as an award!

Student/Parent reactions

It is important for people to be a part of the learning process. Both parents and children appreciate the opportunity to share their impression of the course and make suggestions for future classes. Participants will enjoy capturing suggestions on the board, discussing them, and making records of them for the future.

Homework Puzzles

Play board games and puzzles with your friends and family.

Some advice for parents

"Become aware of what you yourself are capable of before you attempt to outline the rights and responsibilities of children.

First and foremost you must realize that you too are a child, whom you must first get to know, to bring up and to educate."
— Janusz Korczak

Informal family education

Many parents express feelings of uneasiness or even panic when they discuss mathematics with children. Make problem solving as informal as possible, so you can enjoy it together as a family.

Facilitator vs. teacher

The math circle leader is a facilitator rather than a teacher. One of the major differences is that the teacher is considered an authority figure, while the facilitator is considered an equal. The essential trait of a math circle is free discussion between everyone. The role of the facilitator is to ensure that everything goes smoothly.

The main tasks of the facilitator are to help the group:

• Set ground rules and uphold them

• Keep activities focused and moving

• Observe and analyze

• Document ideas

• Reach a consensus

Motivation

Most parents want their kids to have an intrinsic motivation to learn. To increase your child's motivation, consider the following:

• Let children be autonomous. Give them space to explore freely.

• Remove external rewards such as grades. Do not use praise to influence children's behavior.

• Make sure the child is comfortable with their group.

• Establish clear goals. Why do we need this? What are we learning?

• Help children recognize each others' contributions.

• Encourage children to share their knowledge and organize community events, such as math contests and festivals.

Setting up

Space organization

Young kids love moving to new spaces for new activities. Novelty stimulates creativity and problem solving. It is helpful to have a big rug to sit on. Switch to a table for drawing and puzzles.

Some kids need to sit still. Others need swiveling chairs or exercise balls so they can wiggle. When kids know they sit on a rug for stories, stand in a circle for hand games, and gather around the table for puzzles, transitions between activities are faster and smoother.

Movement

Explain to the children that we humans get distracted and may feel threatened by sudden, fast movements, especially in our immediate space. Suggest that kids move slowly and keep their distance from others to keep everyone focused.

Snack

Thinking and concentrating requires a lot of energy, and children get exhausted quickly. A short snack provides a much-needed break. If your class lasts for more than an hour, having snacks is highly recommended. You may request that the parents give children an extra snack, and set-up a special area that is quick and easy to clean.

Grown-up participants

Here are some ways that grown ups can participate in your math circle (other than leading):

The photojournalist takes close-ups of matchstick creations, candid pictures of kids arguing about paradoxes, and group portraits with everyone holding their pictures of infinity. Ask everyone in the group for written permission to take photos.

The storyteller can type up a live report on a laptop, or take notes for later. Good stories are human: they focus on the kids. They capture the excitement of finally finding that last number in a pattern, the hard work of checking all the possibilities, or the healthy frustration of being stuck in a problem. The stories can later be posted online for everyone to read and discuss.

The data collector can write down the most surprising things kids say, or chart who worked with whom on a project. This can answer your questions and worries, such as, "Do I really address girls more than boys?" or "Do kids concentrate better on bouncy balls or chairs?"

The tech assistant helps kids to spell words, to glue or fold paper, to hold a ruler still, and to read problems. They help kids do what they want, without taking over their adventures.

The mediator listens to kids if there is a conflict, and helps all sides to tell their stories. Sometimes a kid may be sad, and needs a dedicated adult for company. This is a nurturing role for an active listener.

The inspirers usually sit aside but close, in pairs or small groups. It's a perfect role for someone who mostly wants to observe. Inspirers chat among themselves (quietly) and build a big tiling puzzle, draw a six-way symmetrical picture, or come up with the funniest paradox. The driving force behind what the inspirers do is that they — follow what the children are doing. Kids peek at these more complex, more grown-up versions of their own tasks, and become inspired. Co-working with interested adults creates a good flow.

Appendix A: Debate rules

Model peaceful conflict resolution. Pull team numbers from a hat to determine the order of team presentations.

Counterbalance the tension of competition with the comfort of choice. Let students pick any problem from the list. They can either work on several problems, assigning them to different group members, or work on one problem as a team. There is no need to solve all problems. After a set amount of time (half an hour or so) announce the end of work within teams, and the beginning of a group discussion.

The solution becomes stronger when everyone in a group has a stake in it. Teams take turns presenting the results within a time limit. After a solution is presented, the team members have a minute or so to either accept the solution as a team, or make a last moment change or addition.

The presentation may uncover misunderstandings between team members or bring last moment breakthroughs. If a team member dissents from the team decision at the point of presentation, the team can decide to forfeit the problem (and get zero points as a result). A solution cannot receive points unless all team members accept it.

Discussing the problem with other teams improves the solution and adds new perspectives to it. Other teams have good reasons to watch presentations carefully! After each presentation, other teams will be given a chance to respond. Each team's response should take no more than couple of minutes. If a responding team finds an error, they get an extra point, and the presenting team loses a point, in addition to not gaining the points assigned to the problem. If a responding team has a different solution, they are awarded full points assigned to that problem. It is possible for a team to solve just one problem correctly and win the contest by finding mistakes in opponents' solutions.

A parent volunteer may be asked to keep score; otherwise, the teacher keeps score on the board. The teacher facilitates the discussion, maintaining timing, order of presentations, and problem scoring in situations that are not clear (e.g., partial credit). The teacher does not approve or disapprove of the teams' decisions, fostering the autonomy of the students.

The debate may be over-exciting to some children, leading to distraction and loss of information. Bonus points for teams may help with "keeping cool" and preserving supportive, scientific atmosphere. The team gets bonus points for:

• Carefully and quietly listening throughout the entirety of each presentation (encourage children to draw or write on-topic notes to prepare their responses)

• Talking and moving quietly, without sudden loud noises or jumps, to avoid distracting others

• Being friendly, kind, and logical when talking to judges and other teams

Appendix B: Stuck?

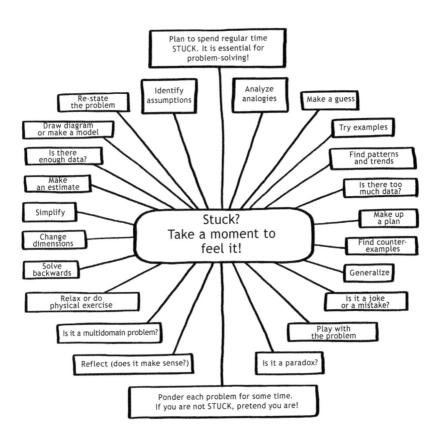

Appendix C: Parent Bingo

PARENT BINGO:

	A	B
1	...Bravely interact with software, manipulatives or other systems, having no clue about the purpose?	...Answer questions in math actions rather than words?
2	...Drop whatever they are doing and get totally distracted after a significant "A-ha!" moment?	...Notice mathematical beauty and share it?
3	...Have to "just freely play with the new stuff" before paying any attention to suggested activities?	...Notice and celebrate a math success, big or small: jump exuberantly, high-five a friend, happily yell, run around, grudgingly say, "I guess it was OK"?
4	...Have weird difficulties moving from objects to pictures, or from pictures to symbols, toward more abstract representations?	...Notice a pattern or any general rule, no matter how strange and silly?
5	...Persevere, maintain attention, concentrate, and cope when the going is tough?	...Create a little ritual, repeating a phrase, a song or a gesture in the same place in the activity?

Do you notice how kids...

C	D	E
...Find a loophole in questions, problems, statements?	...Ask "what if?.." questions, however silly or fantastical?	...Question whether something that seems right (e.g., comes from an authority) is right?
...Find the same idea in two or more different contexts?	...Explain an idea to someone, correctly or not?	...Put a wrong idea to some use?
...Evaluate a task in any way: "This is a hard problem" or "I like this project" or "I don't want to do *that* many computations in my whole life"?	...Switch from pictures to objects, from words to symbols, or between some other systems of representations?	...Experiment with examples, objects, pictures to find out if an idea is right?
...Are fascinated by ideas far above their level of competence, or amused by ideas far below?	...Come up with their own original ways of doing something, right or wrong?	... Anticipate an operation result, a graph shape, the next object in a pattern, etc. before the math is carried out?
...Notice a mistake and try to self-correct?	...Significantly change their math behavior when in a group?	...Hold their own in a discussion, argue, explain ideas to others, share guesses?

Appendix D: Resources

Books

For more information on math circle logistics, as well math circle lesson plans for middle-school kids, I highly recommend the book, *Mathematical Circle Diaries, Year 1: Complete Curriculum for Grades 5 to 7* by Anna Burago. *Moebius Noodles* by Maria Droujkova and Yelena McManaman is chock-full of fun and profound activities for younger crowds. *Mathematical Thinking* by John Mason is a great resource for dealing with being stuck in problem solving for older kids.

For more suggestions, please visit the Moebius Noodles website (moebiusnoodles.com) and look out for more Delta Stream Media books.

Online resources:

Communities

- Moebius Noodles
- National Association of Math Circles
- Art of Problem Solving
- Math Teachers' Circle

Games, materials and online manipulatives

- NRICH

- Puzzles.com

- Cut-the-Knot

- Math Pickle Wolfram Demos

- GeoGebra.org

- National Library of Virtual Manipulatives

- Tricky Pre-K Math

- Beast Academy

Extra resources per topic

Topic	Extra resources
Introduction to thinking	*Habits of a Systems Thinker watersfoundation.org/systems-thinking/ habits-of-a-systems-thinker/*
Strange Statements	*"The Game of Logic"* by Lewis Carroll *http://www.gutenberg.org/files/4763/4763-h/4763-h.htm*
Tiling Puzzles	Cathedral, or Blokus Trigon board game or online *pentolla.com* More pentomino puzzles: *scholastic.com/blueballiett/games/ pentominoes_game.htm banjen.com/windows/en/pentamino*
Problem Debates	*en.wikipedia.org/wiki/Zeno%27s_paradoxes*

Topic	Extra resources
Infinity	PBS NOVA Fractals: *youtube.com/watch?v=LemPnZn54Kw* *"One, two, three.. infinity "* by George Gamow
Symmetry	Mathisfun symmetry games: *mathsisfun.com/geometry/symmetry-artist.html* *Mirror images: Masters of Deception: Escher, Dali & the Artists of Optical Illusion* by Al Seckel
Lateral Solutions	River-crossing plank puzzles *puzzles.com/rivercrossing/play.htm* *clickmazes.com/planks/ixplanks.htm* Make a maze: *mazegenerator.net*
Game Festival	The World Cafe set-up principles: *theworldcafe.com/pdfs/cafetogo.pdf* Online board games and puzzles: *schooltimegames.com/Mathematics.html* Board game printouts: *thinkfun.com/familyplay/strategy* *puzzles.com/PuzzlePlayground/PencilNPaper.htm*

Summary

Thank you for taking time to read this book. Here I would like to emphasize some of the main points one more time.

• Young children are capable of deep thinking, especially if adults help with capturing ideas, computation, and other task management and tech support.

• Insight problems and deep concepts ignite interest and motivate young kids to ponder.

• Taking one's time to ponder a problem is essential for the development of deep thinking.

• Making, noticing, collecting, and resolving mistakes should be encouraged and modeled by teachers as a fundamental skill.

• Deep thinking is beautiful and awesome. If it does not feel so overall, you are doing it wrong. It is also dangerous and frustrating at times, like an extreme sport, and requires skills and community support to survive.

I wish you and your children and students all the best in your problem-solving adventures. If you have questions, please connect with us in the Moebius Noodles Questions and Answers Hub (*ask.moebiusnoodles.com*).

Best regards,

Julia Brodsky